The Love Makers

Aifric Campbell is the author of three novels: *On the Floor*, longlisted for the International Orange Prize 2012 (now the Women's Prize for Fiction), *The Loss Adjustor*, and *The Semantics of Murder*. Born in Dublin, she studied in Sweden and spent fourteen years as an investment banker at Morgan Stanley where she became the first woman managing director on the London trading floor. Her writing has appeared in the *Guardian*, *Wall Street Journal*, *Irish Times*, *Sunday Telegraph*, and others. She was awarded a research fellowship at UCLA and residencies in Europe and the United States. She received her PhD from the University of East Anglia and teaches at Imperial College London.

The Love Makers

A Novel and Contributor Essays on the Social Impact of Artificial Intelligence and Robotics

Author and general editor:
Aifric Campbell

Scarlett and Gurl, a novel by Aifric Campbell

Contributors:
Ronny Bogani, Joanna J. Bryson, Julie Carpenter, Stephen Cave, Anita Chandran, Peter R. N. Childs, Kate Devlin, Kanta Dihal, Mary Flanagan, Margaret Rhee, Amanda Sharkey, Roberto Trotta, E. R. Truitt, and Richard Watson

Goldsmiths
Press

Contents

Part I

The Novel

Scarlett and Gurl

Aifric Campbell

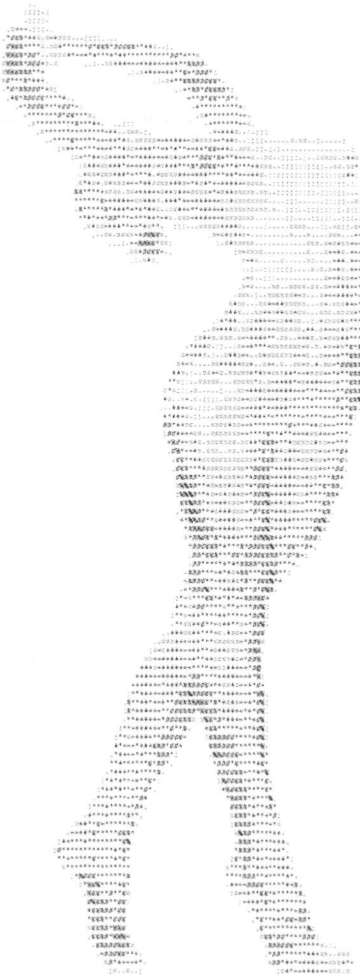

My soul is wrought to sing of forms transformed to bodies new
and strange!

(Ovid, *Metamorphoses 1.1* (C.E. 8),
trans. Anthony S. Kline, 2000)

You can know no more about the future, I was recently assured
by a friend, than you can know which way a kitten will jump next.
(H. G. Wells, *The Discovery of the Future*,
a discourse delivered to the Royal Institution, 24 January 1902)

What is the future of feeling?
(J. D. Bernal, *The World, the Flesh and the Devil*, 1929)

This story starts

a long, long time ago – two thousand years ago, in fact – when
a Roman poet called Ovid wrote a story called 'Pygmalion'. Like
all great stories, it's about love and loss, about a sculptor who
*carved a figure out of snow-white ivory and fell in love with his
own creation.* Now this was true desire, real lust: *passion for
this bodily image consumed his heart, he kissed it and felt his
kisses returned.* Pygmalion wanted the statue as his bride and
the gods granted his wish: *he touched her breast with his hand,
the ivory yielded to his touch and lost its hardness, she blushed
and raised her bashful eyes.* So, the happy couple married and
had a baby.

Four years ago, Ovid sent me on a road trip through the
forest and into the future. I have seen titanium fingers heal
the human heart, lovers locked in silicone rapture, babies
slumbering in a steel embrace. I have spoken with the Makers
who realise our ancient dreams of speed, longevity, and love;
who craft the maths and mechanics to gift us bodies new and
strange: stronger, faster, and surely better! It was the strangest
journey I have ever made, for when I reached the end I saw the
future has already arrived. There will be no going back to the
old ways. We have shed our skins for a technological enchantment. We have changed the power of dream.

Oh, how the poet would revel in these changing times!
What Pygmalion had to beg from the gods can now be
delivered by human ingenuity. Today, in Lab, with a pale sun
slipping out of sight, the chat was of catastrophic forgetting.
About previously learned responses that are suddenly lost.
This is a problem for machines, not for humans – at least,
not yet. But it's coming. So happy two thousandth birthday,

Ovid. People say that Shakespeare got his ideas from you. People also say that Freud got *his* ideas from Shakespeare. But that doesn't have to matter anymore. What matters is that love itself – enduring, eternal human love – will be forever transformed.

Aifric Campbell
www.aifriccampbell.com

DAY

It started up

thirty-eight minutes ago as they pulled out of the city. A dry sobbing somewhere in the middle of the carriage, the kind of crying that speaks of an awful thing that can't be borne. The kind of misery that could leach into your life if you linger. The kind of bad news you never want to know. People sit down and then sneak away like it could be contagious. But how do you leave someone alone with that pain?

She rises quickly from her seat, then hesitates, because this is exactly how trouble starts. *You bring it on yourself, getting messed up in other peoples' lives*, is what would Frank say. If he was here and not three thousand miles away and asleep, a storybook on his chest with Fintan sprawled beside him, face down like a swimmer in his Batman pyjamas. She steadies herself against the carriage sway. There's nothing at all about this that calls Fintan to mind, except perhaps the crying loop that a three-year-old slips into when the only cure is a big, warm hug. So it's that, the feel of him hot and snuzzling at her shoulder, that propels her towards 29D to ask,

'Are you OK?'

A girl, a young woman in a pink puffer, hands flopped like chicken pieces on her lap. Who doesn't answer, or even turn her head. Just shudders, staring out the window like the source of her grief is hidden in the dark forest flying past. The puffer is a vile cerise. Her hair's a day off washing, scraped back, and clamped tight in a scrunchie, a thin brown watermark pushing out the blonde.

At least I asked, she thinks, backing away to her own seat and her own business. She taps her earbuds, but even at maximum volume she can still sense the sobbing beneath the bass.

The Buick glitters like a giant golden bauble in the deserted parking lot and, for a moment, she wonders if this is some bizarre prank. She glances quickly round but there's nothing, just the vanishing rear of the train, four walls of massive conifer, and a sweep of empty blacktop beneath the leaden sky.

'Frank, you've got to check out my rental!' She scans her phone over the pristine white tyres and voluptuous engineering.

'I've been trying to reach you.' Frank's voice is tetchy. He's not interested, not listening because it's day three now and the nanny is still sick, Fintan is still weepy, and work undone weighs heavy on his shoulders.

'You know how the signal is, out here in the wilds.'

'So, are you done?'

'Just heading to Lab now.' Cold snips at her fingertips as she taps in the lock code and the Buick's headlamps wink suggestively.

'I thought you were there already!'

'I had an unscheduled meeting in the city.'

'Have you made up your mind? Are you yes or no?'

'I'm still leaning to no.'

'Je-sus.' She hears his weary frustration. 'We've been over this a million times. There's a bid on the table and both your partners want to sign. There is no logical basis for you to say no.'

'I've just been to meet the new buyers and I don't trust them.'

'Trust doesn't come into it. This is a takeover, not a marriage.'

'It's not that simple, Frank.'

'It's precisely that simple, but you're turning it into one against two.'

'We are three equal partners.'

'Bullshit. No one's equal to Felix. And Colin is a definite yes. Hang on – Fintan sweetheart, the blue paint is right there, see? Look, it's Christmas Eve and there's only one flight that gets you home.'

'And I'll be on it.' She yanks at the car door.

'You know this deal is going to happen. You're just being – '

'What? What am I just *being*, Frank?'

'You don't want to let go. I know you've spent years on this but – '

'Great. So now I have my partners *and* my husband telling me what to do.'

The clunk is muted. Inside, the Buick La Salle is a sound chamber that insulates from the external world. It's a creamy vanilla excess of pillowed leather, gold stitching, walnut dash. So this is what she gets for insisting on a drivable disconnected, as if she was some kind of vintage hysteric. Of course, the retro look is surface level: the sensors have already registered her presence with a velvet hoosh of humidified air. She has only to speak and the Buick will limber up for departure. Most disappointing of all is that the windscreen is already busily recasting her worldview to the sunshine default, standard issue now ever since the clouds came to stay. She snaps the override button and kills the internet connection. The yellow filter bleeds from the glass and the world returns to its grey familiar.

'Sorry, Frank. I know you're up to your eyes without her.'

'You know what the solution is.'

'And you know my answer is no.' The silence stings. She pictures Frank sucking in his lower lip, holding back a barb.

'So tell me,' she says lightly, 'what's Fintan up to?'

'He's painting a "Welcome home, Mummy" sign.'

'Ah,' she strokes the white wheel, pictures paint blotches, a watery blue *M*.

'It's a surprise.'

'Then don't tell me!' she smiles, transatlantically. 'I don't want to have to pretend surprise – I want the thrill of actually feeling it.'

'You asked what he was doing.'

'Pink puffer,' she murmurs.

'What?'

'Oh, nothing – just a girl who was crying.'

And there she comes, striding past the station's locked entrance, cheeks red with cold, bleached wisps floating free. She's short, very short – maybe 1.55 m, and that's with heels. And very young, or at least young enough to be wearing a short black dress with lacy frills in minus two.

'You still there?'

'How's Fintan's tooth?'

'Gum's still red.'

Pink Puffer stops at the top of the steps, scowls at the parking lot. Twitches her nose like she's picked up a scent, but there's nothing to smell in the clear air, crisp and unwelcoming in this northern pitstop.

'You're giving him the ice chips? I hate to think of him – '

'Missing you.'

'Having a sore tooth, I meant.'

'He hates you being away and it's getting worse. He cries all night.'

'I wish I was there.'

'No you don't.'

'How can you say that!'

'You know I didn't mean it that way.' Frank's sigh is long and heavy. Warmth, now, is what's needed to cross this hump. She closes her eyes, pictures the Christmas tree, Fintan in his Santa suit, Buster with a tinsel collar.

'I *do* wish I was there.'

'If you did, you would be.'

'Fucksake, Frank.' She smacks the wheel.

'Hey, that's not an accusation, just a statement of fact. If you want to keep up this unnecessary travelling, you know what we need to do.'

'Let me talk to Fintan, I'll talk to him now.'

'You know that just sets him off.'

Of course, Frank is right. Voice makes her present, makes Fintan feel like she's just down the road and not on the other side of the globe. The more she talks the worse he gets, then Frank has to take the phone and it's all *hurrybye love you* and a severed line.

'Frankisalwaysright', she whispers, lips close to the microphone, and she knows he's smiling at this running joke between them, the one with the hard crystal of truth at its core.

'Kiss him from me. I'll call later for his bedtime story.'

'Good luck with Colin and Felix. Don't make it harder on yourself.'

And Frank is gone. She docks the phone on the dash like a totem and stares at the black screen. Her hand hovers in the space between home and wheel – she should call back right now and tell him she can't wait to get back, tell him she misses him. But it's so hard to find the words, and the truth is that lately she misses Frank most when they are together. It's their shared past she longs for, the one they cannot seem to build on.

She reaches for the gear and rolls towards the exit gap in the forest wall. Pink Puffer turns, eyes cruising over the Buick's hips, and it's a feral look in the rear view, her steady gaze tracking the golden chariot as it swings out on to the empty freeway and disappears.

As soon as her voiceprint activates the steel door she is hijacked by a wave of nostalgia. It is seven years since they bought this underground lab from a tech start-up, after the CEO slit his wrists and locked himself in the vault so he could bleed to death online. 9,251,102 candle-flame emojis kept YouTube vigil while the celebrity coder transitioned from man to myth, as if his final act – the wordless blood-letting and his corpse propped up against the server cage – carried an important message, rather than the consequence of a lethal cocktail of cashflow crisis and burnout. Techworld went into a tailspin about bad karma, which meant they could secure the lease on this high-spec lab for a song. For all their brilliance, she has come to realise that programmers are a superstitious lot, like peasants in a hobbit world of rumour and conspiracy that is fuelled by daily newsfeeds about cybercrime. So the partners are careful to nurture their craftsmen and keep them safe: everyone is nanochipped so that real-time biofeeds can be monitored for emotional tur-bulence and whereabouts.

The lab is silent and chilly as a church. Five coders sit cloistered by a massive, curved screen wall, like a cluster of wizards inside an enchanted fortress. Lines of white and green code slither across the black, like magic spells con-juring up a future that is elegant, robust, and efficient. She steps closer, her movement registers on the monitor wall like an interloper. But no one says hello and she does not speak. Everyone knows she is here, but physical presence is considered an eccentricity. What's the point of face to face? Why fly with all the security hassle? Old-school, says Colin, but they have both worked together for so long now that he tolerates her quirks. And anyway, she has always loved the shop floor, the quiet fury of concentration where work unfolds in a sacred hush with occasional bursts of banter. She smiles now at the seasonal gestures – Kale sports a Santa hat, Xiang has clipped antlers to the back of his chair. Even the Lab mascot suspended from the ceiling has had a makeover: the large furry fly with electrode spikes in his cranium has grown gossamer angel wings.

Xiang and Kale stand side by side, arms folded, watching a web of yellow vectors converge on-screen. All important work is silent now, since math makes our world. Human industry grows ever more hushed and she sometimes longs for the noisy collaboration of the old days. The echo chamber of idea exchange, the physicality, the yelling and shouting of her old life on the trading floor where you could reach out and touch a living being. Another reason she doesn't want the deal: she will lose the few tangibles they have.

She turns away from Code and heads down the corridor, tracking her shadowy avatar on the monitor wall. This lab that Felix bankrolled, that she and Colin built, is her home-away-from-home, and now she has reached the end of a wonderful adventure. Her life has been bookmarked by deals: lose one and find another. And the post-deal emptiness brings an adrenaline crash – another petty reason to keep resisting. Or maybe Frank is right. Maybe she really is just afraid to let go?

FlyBoy sits cross-legged behind the glass wall of Test. Slender and barefoot in shorts and T-shirt, he crouches like a child assembling his toys on the floor. Skinny arms and legs spattered with freckles and a tumble of strawberry hair that blazes against the lime-green walls. He still looks like a schoolkid and has not aged a day in the seven years since she found him. At just

158 cm and 53 kg, he is a perfect candidate for recombinant HGH treatment, with a predicted 95 per cent success that could take him to 166 cm in 28 months. She has discussed the idea with him, but he does not care. He rarely ventures out: in fact, he'd prefer to live here all the time, but they only let him bunk over every third night in Sleep, where he spends a few hours suspended in a slothlike hammock. Otherwise he is shuttled between Lab and his safe house, chipped and tracked like everyone else.

FlyBoy twists round suddenly, as if he senses she's watching, and waves a slow side to side that could be hello or goodbye. His blue irises are clear as the day she bumped into him on campus, smacked his nose on her collarbone, and dropped his lab box on the grass.

'You're pretty tall,' he muttered.

'And you're a little boy who doesn't look where he's going.' He was already down on his knees bent over the plastic box that flickered and buzzed.

'What's that?'

'My research.'

'Insects for your school project? And by the way, where's my apology?'

' "Collision Avoidance and Escape Behaviour in Blowflies," ' he looked up. 'My paper was in last month's *Nature*.'

'So you're not a schoolboy.'

'I am a twenty-four-year-old bioengineer. And I apologise for walking into you.'

'Tell me about the box.'

'Why do you want to know?'

'Because I'm looking for a good story that needs money to come true.'

They settled on the grass with the white box in between them while students swarmed across the quad.

'When I was a kid,' he began, 'my mom used to get up at six to make fresh pasta for a deli. By 7 a.m. it was already twenty-eight degrees and, soon as I could walk, my job was to zap the flies. I have this little red plastic swatter, I get *this* close and WHACK – but they're off, buzzing like crazy round the kitchen. They swoop and dive and twist and switch direction, wings flapping at 150 beats per second. It's boy versus fly and my kill count is close to zero. But all I can think about is why don't they kill themselves? You ever see a fly in a head-on collision? Ever wondered how they don't just go splat into the wall? How an insect with a brain the size of a pinhead beats every single flying machine that humans have ever invented?'

He leant towards her, crossing his palms on the box. 'Consider the fly's navigational system. Think about how much processing and calculation is going on there. What are the neuronal control systems that allow a fly to outsmart a jet fighter? *That* is all I ever wanted to know. Because if you can understand the fly, you can revolutionise flight. You can design a completely new flying object. Planes don't have to look like raptors. Jet fighters are unmanned.'

'So what do you need to get it done?'

'What every scientist needs – funds and freedom. Then I can get on with my work and not waste my time writing research.'

FlyBoy slips his right hand in the pocket of his shorts and withdraws a clenched fist. Slowly, gracefully, his long white fingers unfold to reveal a glossy ivory sphere. He raises his arm, holds it delicately between thumb and index finger as if he is admiring a giant pearl. And she is struck, as always, by its satin sheen. For Volo is made from the toughest material on earth: silk fibres harvested from Darwin's bark spider. And at 520MJ/m^3, stronger than steel, tougher than a bulletproof vest.

FlyBoy places Volo on the floor and nudges it with his toe. It spins away, shoots a high-speed diagonal towards the far wall, then swoops to a sudden landing in the corner. He bends, places a second Volo on the ground, wiggles his toe, and it swerves round his foot to land on the wall behind him, crawls upwards, and darts back to ground. FlyBoy stands up and begins a slow walk around the perimeter, hands in pockets. In the long months of beta testing his body was purple with bruises, but now he strolls unflinching through Volo's flight path.

He has unpicked the mystery. And together they have built this prototype: a multisensory micro air vehicle that outperforms a housefly and weighs no more than a nectarine. Volo will make the jet fighter look like a dinosaur. What's more, it flies itself: it is completely autonomous – no human intervention required.

And now it is for sale: FlyBoy and the intellectual property, the project that Felix funded, the math model that she and Colin built, the wizards they assembled, the code that created Volo, the engineering that gave him life. The bid is on the table. Her partners are ready to sign. But still, she hesitates.

Her problem is an old one: what humans do with all the clever things we invent. She does not trust the buyer. She does not like FlyBoy's new handlers, whom she startled this morning with her impromptu visit. The bloated corporate whale that has spent the last decade gobbling up artificial intelligence just because it can. Like the nursery brat who hoards all the toys just to stop others from having them. No clear strategy but plenty of dosh – they are fat and woozy with excess capital. She has studied their acquisition trail and she knows their dreams: drones that make up their own minds. No human dithering to fuck it up.

FlyBoy looks up and grins. It's a rare moment, like the gift of a child's smile. She presses her fingertips against the glass. He holds up a splayed palm, adds a thumb and forefinger, and she laughs aloud.

'Yes, seven!' she says. Seven whole years at Lab, and she feels the prick of tears for the wonderful adventure that will soon be past tense. The new parents will arrive and FlyBoy will live on here with the wizards and the furry mascot, the partners will say goodbye, and the separation will be complete. Her heart rate rises, her body temperature elevates. She runs her finger over the fleshy base of her thumb – sometimes she swears she can feel the chip beneath her skin, though of course that's ridiculous – it's no bigger than a hair's width. The tiny polymer parasite that feeds off the drama of her daily life and lays her bare on-screen so her emotional arousal can be examined by her partners. In reality, it's Colin who obsesses over the biofeeds – heart rate, blood pressure, endocrine function, toxicology – a comprehensive neurophysiological newsfeed that identifies problems and tracks performance. But while the data reveals the symptoms, it cannot reveal the cause. Motive remains the only secret.

'I know why you pulled that stunt this morning.' Colin draws level with the glass wall, but he doesn't look at her – just stares into Test. 'You disabled your chip for a couple of hours so I wouldn't know what you were up to. Do you have any idea what I was thinking when we couldn't track you?'

'Let me guess: that I'd become today's cybercrime headline? "Mum Kidnapped in Algosnatch."'

'Not funny.'

'So you were picturing me tied to a chair in some dungeon lab while Volo's algorithms are beaten out of me?' She grins. 'I'm surprised at you Colin, you're turning into a hobbit.'

'There's been five high-profile algosnatches in the last six months, all of them within a hundred-mile radius,' he turns to face her. 'Volo is the hottest property in town, which means *all* of us are potential targets. What's the point in having implants if you are going to hot-wire them? Plus, we agreed, remember? You, me, the coders, and him,' he jabs the glass as FlyBoy strolls past. '*Everyone* is chipped so we can always be tracked.'

'Felix isn't chipped.'

'That's because Felix never goes anywhere. *And* he doesn't touch the code. *And* he's on the other side of the planet. But now, just as we're about to sign this deal, *you* decide to go AWOL?'

'So it was the *deal* you were worried about, not your old partner.' She touches Colin's shirtsleeve and he flinches. At the charge or the touch, she cannot be sure. Here, now, in the dry prickle, she can feel the static in her hair.

'I just wanted some privacy.' She heads to the watercooler and reaches for the pressure gauge to discharge.

'Bullshit. I know you went to see the buyer because I went into the code and hooked you up again,' Colin follows behind. 'And by the way, you're not as smart as you used to be. I was able to overwrite you.'

'Gold star for you, then,' she peers into the fruit bowl. 'Bananas and grapes – Jesus, Colin, is that the best you can do? I'm starving.'

'I'm in training for a triathlon.'

'You know,' she peels the banana, shaking her head, 'sometimes I just can't believe it's still you.'

Colin twitches his neck – the past makes him itchy. But it's still a shock to see the athlete he's become. To think of the blubbery shambles he was when they first met: sugar-coated fingers stabbing at the keyboard. Twenty-seven kilograms and total hair loss marks their thirteen collegiate years, and Colin's fat suit has melted away to reveal a trim, lean string of muscle that can hold her gaze without blinking. Though not right now, because he's angry: she sees it in the set of his jaw – old frustrations rising to the surface. But, despite the squabbling, theirs is a near-perfect working relationship, and she could not be more surprised at the journey from reluctant beginnings to operational excellence.

'Time to talk to Felix,' he turns away, 'and I'm going to tell him what you've been up to.'

'Snitch.' She dumps the peel and follows him to Hub.

Colin stands in front of her, arms folded, admiring the animated explosion of colour: a giant blue and black cochlear spiral that expands and contracts as if the wall itself was regurgitating. She steadies herself against the table edge, tries to focus on his stubbled skull, and ignore his wife's latest installation. For Colin, it was love at first sight: he was captivated by the synchronised-swimmer-turned-fractal-artist and her infinite iterations, and she in turn was overjoyed to finally find someone who appreciated the beauty of algorithmic art. All the plotlines in his life have since converged in a perfect resolution of personal and professional. Their twin daughters could code before they could read. Gender and number decided before conception. Colin did not want a son; he simply wanted to reproduce more variants of his wife.

The spiral swells and then implodes into a black tunnel that grows ever deeper. She blinks at a huge orange petal that

yawns and multiplies, and threatens to pitch her into a ver-
tiginous abstraction that will swallow her whole.

'Switch it off, Colin, before I puke my guts up.' He snaps the
remote and blanks the wall to white.

'Sorry.'

'It doesn't matter.'

'Your wife's art just makes me dizzy. But that doesn't mean
I think it's – '

'Doesn't matter what you think.' And it doesn't, which is
precisely what she loves about Colin – his predictable indiffer-
ence to sentiment.

Hub's white wall crumbles into a shimmering illumination
and, slowly, the rock city emerges as a jigsaw of twinkling glass
structures. Yellow freeways like landing strips, red lights, green
lights, blue lights, and dazzling white floodlights blaze against
a navy nightscape.

'Felix, how handsome you are!' She steps closer; the harbour water glimmers.

'My dear,' his voice behind the panoramic glow fills the
room, 'it's such a very long time since I had the pleasure of your
company on this island.'

'I was just thinking – it's nearly four years since I saw you
in the flesh.' She touches the screen and zooms in to a close-up
of ripples blackly glistening. Recalls clammy nights, waterfront
smells, the blister of margaritas, humidity trickling down her
back. 'Oh, I do miss you, Felix,' she sighs. 'How are you?'

'I am older.'

'I picture you as a distinguished grey. I suppose there's no
point in asking you to break with tradition and turn the camera
on yourself. Imagine, we've been partners now for eight years
and Colin has never once seen your face.'

'Doesn't bother me,' Colin shrugs, leaning against the far wall.

'Of course it doesn't!' she smiles. 'You know, you two really are the perfect match – the recluse and the geek. Where would you be without me!'

Her partners are two fixed points on either side of the world and she is the moving part. From the outset, Felix had insisted on the direct-line arrangement they'd had years before, when she was his banker: he would only speak to her. They make an unlikely threesome – the geek, the banker, and the hedge-fund philosopher – but it's turned out to be a happy operational marriage, and a harmonious journey from idea to execution. Everything that has happened in their time together has been good and positive and brought them closer. What she likes with these two is the comfort of a lingering past that remains unvisited. They do not go back, they remain rooted in the present and a theoretical future. And they have worked together for so long now that she cannot imagine it otherwise.

'I see your neighbours are being remodelled, up there on the peak,' she swipes away from the harbour to the mouth of a vast black hillside crater where a platoon of yellow dirtbots scurry about like ants.

'Yes,' says Felix. 'Change is the comforting constant.'

'Don't you ever think of leaving? Following the East–West drift?'

'And where would my home be now, do you think?'

'Another planet, Felix. Where it always should have been.'

'I hate to interrupt this cosy catch-up, people,' says Colin, 'but can we save it for the after-party?'

'You look a little peaky, my dear,' Felix ignores the interruption. 'I noticed your biofeeds were out of action for a while.'

'That's because she switched off her chip.' Colin steps closer to the screen. 'So she could cover her tracks and sneak off for a surprise visit to our buyer this morning.'

'Haven't you two got something better to do than ogle my vital signs?'

'So you've been interviewing the new parents?'

'Call me old-fashioned, Felix, but I still believe in face to face. Which is why I am here – risking Christmas with my little boy.'

'They said she was abrasive,' Colin addresses the screen wall. 'Like she doesn't trust them.'

'I don't.'

'They said she was insisting on an ethics committee.'

Felix chuckles faintly. She pictures his thin-lipped twitch, sitting in the shadows behind his desk, high above the city's gleam.

'I don't like what they could do with Volo,' she says. 'Every single one of our buyer's key acquisitions in the last two years has had a military application.'

'Christ, I don't believe I'm still hearing this crap! I mean, how many times have we been over this?' Colin presses both frustrated fists against his temples. 'There are *tons* of possible applications for Volo – weapons, surveillance, blah, blah, blah – but you insist on ignoring all the incredible life-*saving* applications. Like disaster response, for example. But either way, it's not our problem. We are selling an AI – we are selling *potential*.'

'First, do no harm.' She watches Colin pace back and forth.

'So now you want a Hippocratic oath written into the sale contract!'

'And you just want to sell.'

'Everything is always for sale.' Colin stops dead in front her. 'Wasn't that your catchphrase back on the trading floor?'

'My dear,' Felix interrupts, 'did you really expect that we would arrive at a point where money didn't talk?'

She steps up to the flickering city. 'Remember Feynman? Crouched in his truck in the desert, watching the mushroom cloud. "Scientific knowledge is an enabling power to do either good or bad, but it does not carry instructions on how to use it." '

'Here she goes again,' Colin flings up his hands. 'I refuse to keep having the same conversation.'

'So neither of you are the least bit concerned about Volo being turned into a drone that makes its own decisions? Tell me, then – at what stage are we accountable?'

'We're not!' Colin spins to face her. 'Look, if I sell you a car and you decide to mow down an entire family out for a Sunday walk, am I to blame?'

'Remember Oppenheimer's remorse.' She palms the screen, zooms in and out on the water because she knows it annoys him.

'I don't give a toss about history. And you can stop swiping at the screen just to piss me off. We need a decision. Are you a yes or a no?'

'I am not a yes.'

'So you're a no.' Colin shoves his hand in his pocket. 'Right. OK. Well, I'm a yes. I want to sell Volo.'

She looks away, watches the red flashing light of an approaching helicopter grow closer. Over to the east a fog bank squats malevolently, stealing in slow and steady to smother the city.

'Well then, Felix,' Colin squares up to the screen, 'whose side are you on?'

'I do not take sides. I make informed decisions.'

'You want to sell Volo?'

'I am ready to accept the bid.'

'Two out of three wins.' Colin turns to face her.

'This is not a fucking arm wrestle.'

'Return to first principles,' Felix says softly. 'Capital feeds and nurtures ideas. We incubate, develop, and then we sell. My dear, seven years ago you came to me with a young man and an interesting idea. And we have made the journey from dream to prototype. But we were only ever fostering Volo – there was always an exit strategy. It is not our business to decide on Volo's applications. The future belongs to others.'

'Even though the future could be an autonomous drone that can identify a threat and make its own decisions?'

'Pure speculation,' Colin snaps.

'We have reached the time where Volo must take its first steps,' Felix continues. 'I am neither policeman nor judge. I am simply curious to see what happens. I want to watch the natural unfolding of events.'

'Delegate,' Colin nods approvingly.

'You mean abdicate,' she corrects.

The cloud blanket thickens, eating up the nightscape so that half the city is obscured. A lone glass tower pierces the grubby shroud, its blue light blinking. She shivers, rubs her arms against a sudden chill as if the air in the room is thinning.

'It's like a fucking fridge in here.'

'It's psychological,' Colin mutters.

'Nothing psychological about a fifteen-degree working environment,' she spins away from the screen. 'Anyway, I'm out of here. I've got my last-minute Christmas plane to catch.'

'Yes or no, before you go,' Colin raises his hand.

'I'll call you before I board.'

'Christ,' Colin smacks fist into palm.

'You know, you never used to lose your temper before you had kids.'

'You will call us before 17:00, my dear,' says Felix.

'Hang on a minute – what's going to happen between now and then? All she's doing is driving to the airport.'

'I will be thinking, Colin. You might try it sometime.' She scoops up her bag. 'Bye-bye, Felix,' she salutes the screen. 'It's always a pleasure.'

'You do know that even if you say no,' says Colin, 'we're still a yes.'

'Remind me why I ever agreed to work with you again?'

'Just pointing out that the partnership agreement says "majority vote." '

'Enough bickering!' Felix clicks an invisible finger.

'It's alright, Felix. Colin and I have a long history of bitching. Wait'll you see – any minute now he will give me a big goodbye hug.' She flings her arms wide. He stares, impassive. 'No, really, Colin, you crack me up.'

'Here's your weather report,' Felix adds, and she turns to the screen, the clot of a southbound snowstorm.

'So I need to grab a parka. See how caring he is, Colin? Tell me, Felix, will you always watch over me?' But he does not answer and the harbour is swallowed whole, the city's glitter disintegrates, Hub's wall fades to white and Felix is no more.

'What were you thinking with *that*?' Colin points at the Buick on the security monitors.

'It was the only drivable they had. And, before you ask, yes, I did check the internet connection was disabled, just in case you think I have a horde of car hackers on my case.'

'Sometimes I think you're just deliberately difficult,' he shakes his head. 'Like, for example, your hostility to the biofeeds – I mean why would you *not* want to know your peaks and troughs?'

'Because I like uncertainty.'

'So you deliberately choose to ignore data that could help you to maximise your cognitive performance? You know the coders can't get enough of it.'

'That's because they're control freaks just like you.'

'Doesn't make sense.'

'If you ever read fiction, Colin – which you don't – you'd know that I am what's called a round character. A real person who is capable of surprising you.'

'I have built that into my model.'

'So how does your model compare to the real thing?'

'You are at your most predictable when you are intensely focused. Since Fintan was born you have become less so.'

'Sure you're not straying into mummy stereotyping there?'

'You're more nuanced. It slows things down.'

'Is there a general point you want to make?' She slides open the cupboard and pulls out a green parka with a furry trim on the hood.

'That you just surprise me sometimes. More than you used to.'

'More than when I was a predictable pisshead banker?'

'You're less decisive.'

'It's called growing up and seeing the grey, Colin. The whole point of nuance is that you see more than black and white. I think about the human world, where the ethical issues are the grey. All *you* care about is robustness, elegance, and efficiency.'

The steel door glimmers. All around her is the electrical murmur of process, the hum and sigh and sucking of vented air that dehydrates your eyeballs.

'The landscape gets more complicated,' she continues. 'I mean, you're a parent, too.'

'That's not complicated,' Colin shrugs. 'It's just better.'

'Did you know that Bertrand Russell once said that the happiest men are the men of science, because they're emotionally simple?'

'Anything else you want to say?'

She leans back against the white wall. 'Don't you see what's at stake here? Volo is a game changer. We've developed an AI that will transform lives, that will be a force for good or bad, depending on how it's used. And you're not at all worried about selling it with no restrictions?'

'As I said, you refuse to acknowledge the positives. Like Volo as lifesaver. So no, I'm not worried.'

'And I'm worried that you're not!'

'Time to go.' He is ushering her towards the exit. His eyes behind the rimless glass have a faint greenish tint that she has never previously noticed. But still she lingers; runs a finger along the door. Endings have the dazzling brilliance of closure.

'Love and work, work and love,' she murmurs, 'the cornerstones of our humanness.'

'Is that you or another one of your out-of-date sources?'

'Freud, I think.'

He rolls his eyes, steps towards the door.

'Won't you even miss all of this? And him?' She nods at FlyBoy, crouched down in the centre screen.

'You know that's not how I'm configured.'

'I like to imagine you are still evolving.'

A small twitch of his lips. A virtual smile.

'Tell me something,' she says on impulse. 'How do you feel about your iMom?'

'Feel is not the right word.'

'You know what I mean.'

'It's just as expected: reliable, effective, well designed – and the girls are in the top percentile.'

'They like it?'

'Can't imagine any other life. Part of the family. They whine about not having a gendered design like their friends have, but that wasn't in my spec. Why are you asking?'

'Got to go,' she turns away.

'Still using humancare, aren't you? Even though all the research shows it disadvantages your son.'

'Open,' she tells the control pad, and the doors slide apart. She steps out into the freeze and turns to what could be the last glimpse of her underground home, but Colin is already walking away down the corridor. The door folds into the rock face. The pines bristle overhead; a sharp wind worries at her nose and burrows into her jaw. The Buick glitters like an apparition and she hurries over, hugging the parka, and climbs in.

She stops at the intersection and stares at the dark ribbon of road up ahead. From here to the airport, she knows it so well: the forest corridor, the straight line, the grim press of sky. It has all the feel of a pioneer outback, a scatter of power huts in olive drab with no visible sign of human habitation. And it's only force of habit that holds her at this redundant red light: these days, the traffic is mostly high-security vehicles that ignore signals and the occasional heliscooter whine overhead.

The very first time she came here there was still a lingering population. She remembers a dilapidated supplies store and a man called Colt, who stubbornly resisted the construction of the automated warehouse that would replace him and that squats now on the north-east corner of the intersection, like a fortress in battleship grey.

She looks up at the red light, then down at the phone. She should call Frank, but what to say? Except yes, you were right. And yes, yes, I have always known how this would go in the end.

There's movement in the corner of her eye – a bright blob emerges from around the side of the warehouse. And sure enough it's Pink Puffer, standing straight as a dancer in front of the entrance: hands on hips, glaring fiercely up at the sky as if something about its emptiness offends her. There's no sign of another person, or a waiting vehicle; just a bright red tote on the ground by her feet.

She glides slowly forward on the green, lowers the passenger window, and pulls up level with the stone plinth where the girl now sits, fists jammed in her pockets, head lowered, legs in an outward sprawl like a schoolkid waiting for pickup.

'Are you OK?'

'You asked me that already,' Pink Puffer raises her head. Close up she's maybe twenty: her skin unblemished except for a faint line like a scribble across her forehead; red-rimmed eyes a startling blue.

'You must be frozen out here,' she cuts the engine. 'Do you have a way of getting home?'

Pink Puffer uglies her lips, like the very mention of home disgusts her. 'I was getting a ride. Until I wasn't.' She sniffs and bounces her leg, a pointy kneecap pressing through sheer-black leggings, black suede ankle boots, side-zippered, with a bald shine at the toe.

'How'd you get here from the station?'

'Just started walking,' she shrugs. 'Got off that train at the wrong stop.'

'Why'd you do that?'

'Guess I was feeling kinda moody.' She recrosses her knees and swings her leg upwards like she's kicking away the air. Behind her is some kind of electric crackling emanating from the warehouse interior and a background beeping like a muffled alarm; the patter of invisible footsteps.

'What about a cab?'

'Do I look like I'm made of money? And, anyway, it's too far.'

'I'm on my way to the airport, if you're headed that direction.'

'It's out that way. 'Bout ten miles further north.'

'Where?'

'No place you'd know.' Pink Puffer's fingers probe her left cheek like she's checking for damage, though there's nothing on her white skin but the boned contour.

'Happy to drop you at the airport if you want. But I've got to get a move on now.'

'You normally go round picking up total strangers?'

'You don't look dangerous.'

Pink Puffer hoists the tote on to her shoulder. 'Well, I guess you'll find out!' She stands up, brushes down her frills. 'Hey – Jeez, *only kidding*!' She smiles broadly and reaches for the car door.

'Well, hello there. Nice ride, Miss Buick La Salle. Mm-*hmm*,' she grinds her back against the creamy leather. 'So why'd you pick this baby instead of a driverless?'

'I wanted a real car.'

'So you're a retro girl! You don't want to just lie back and let the world pass by like everyone else? Oh, if I had money I'd just want to lie there and dream.' Pink Puffer runs her fingernail along the gold stitching.

'I like to be actually doing what I'm doing.'

'You mean you like being in charge.'

Scarlett smiles – it's true, her stubborn insistence on continuing to do little things for herself: steering, turning a key.

'You could just watch a movie, take a nap?'

'Not interested.'

'In the movie or lying back?' Pink Puffer grins. 'So what you doing in this dump anyway?'

'Work.'

'Only strangers who ever come here are techies. Though it's usually guys up at those secret labs they got buried in the mountains.'

Her phone trembles in the dock.

'Fintan!' She lunges forward.

'Mummy, come *home*.' His sniffles fill the car.

'Oh, I am coming, Fintan. I will be there very, very soon – '

'COME HOME NOW!'

'One more sleep, Fintan – ' But her answer is lost in his howling; her voice just a scalding reminder of absence. 'Fintan? Frank, you there?' A muffle and phonefall on carpet, Frank's footstep, scuffle, a bang.

'He's OK, he just fell.'

'He fell?' she shouts above the rising wail.

'He tripped, it's nothing, don't worry. Call you back. OK, Fintan, now – '

'Love you,' she yells into the sudden silence, the dead phone; Fintan's cry ghosting through the car.

'Hey,' Pink Puffer turns to her. 'Don't you worry now. Your lil boy got his Frankdaddy.'

'Who told me I'd be home if I wanted to be.'

'Hah!' Pink Puffer snaplaughs. 'He said that? Trust a guy to kick you when you're down.'

'He didn't mean it that way,' she lets her voice trail. Pink Puffer squeezes her shoulder lightly and then withdraws her hand and turns away to the window. She is being the considerate passenger: giving her driver space to recover, pretending there's something worth seeing in the treetops and the grubby cloud.

'Let's go,' she starts the engine. 'So you better tell me your name, since we're going to be stuck in this car for a couple of hours.'

Pink Puffer leans back against the door and folds her arms with a winning smile. A little colour has crept into her white cheeks. 'You can call me Gurl. With a u.'

'You mean G U R L? Is that a nickname?'

'Sorta.'

'OK, well I'm – '

'STOP!' Gurl raises a hand. 'You got to pick a name just for the drive.'

'Why?'

'Cos it's fun!' she shrugs. 'Me and my friend Roxanne, we do it all the time on road trips. Be whoever you want to be.'

'Well I'm fine just being me.'

'Guess I'm gonna have to pick a name for you.' Gurl cocks her head to one side and studies her driver. 'We got the same colouring, you and me. Dead white as a corpse.'

'Except I've got freckles.'

'Bet you burn real easy. Kinda relief now it's always cloudy.'

'It can still get you.'

'Oh, everything can always get you. But let me think now.' Gurl's eyes are blank and clear as Fintan's. A child's frank and curious gaze. 'I got it! Oh, you are gonna love this – it's just the perfect name for you.'

Gurl reaches forward to touch her shoulder. 'Scarlett.'

'Scarlett? Like in *Gone with the Wind*? Though you won't know that movie.'

'Sure I do. My mom was crazy for all the oldies. You got the right hair, black and shiny as a crow. Looks real, too.'

'That's because it is.'

'I know. Hair is my *job*.'

'Do I get one T or two?'

'You get the whole bag. It's Scarlett and Gurl – road trip here we come!'

'It's not a road trip, I'm only giving you a lift.'

'Then why we still sitting on our butts on the side on the road when you got a plane to catch?' and Scarlett rolls back out on to the freeway.

'Where's the online?' Gurl fiddles with her phone.

'I shut it down.'

'You mean you picked a dumb car when you coulda got a smart one? And still had all this deluxe?' She pats the cream leather.

'No connectivity, no tracking. So this car cannot be hacked.'

'Hey, baby,' Gurl's voice is high and sing-song; her whole body slinking in the seat. 'I just got a ride from a nice person headed to the airport. So you can come pick me up there,' she winks at Scarlett. 'I said a *nice* person, hon, ain't no one you know. Just a good Samaritan.' She twirls a strand of hair. 'Oh, I know all about strangers, and trust me, this one is alright. Yeah, I *know* you woulda come Blane but – '

Gurl turns to the window, stroking her thigh with a steady appeasing palm. 'Don't be mad, babe – Roxanne's gonna be fine. I been talking to her the whole time while I was waiting for you. I only started walking when you said – '

Gurl holds the phone at arm's length and Scarlett can hear the tinny anger. Gurl sighs, pulls it back. 'I know, babe. I'm *real* sorry.'

Something flickers as they glide along the forest's edge, a dark squiggle of life at the foot of the trees. Scarlett thinks of the animal names that Fintan loves to hear and repeat – the squirming syllables that lull him to sleep: bobcats, white-tailed deer, long-tailed weasel, wood rats, beavers, and martens. She pictures him from above, asleep on his back, leg flopped to one side.

'You got a good imagination, Blane, I always told you that.' Gurl murmurs low husky secrets into her cupped palm. 'Oh, ain't we a pair,' she lets the phone flop to her lap. 'Your Frankdaddy and my Blane. How is it men always want you home?'

'Frank doesn't.'

'The hell he don't.'

'He's just worried I'll miss my flight. And he's under pressure – looking after Fintan, trying to work, so – '

'Where's your iMom?'

Scarlett frowns, twists her mouth from side to side, keeps her gaze on the road.

'I have a nanny, actually, but she's sick and ...'

'You don't have an iMom? You got all this,' Gurl whirls her arms around the car, 'what are you, crazy? iMoms were *invented* for people like you!'

'Call me old-school, but I still place my trust in human childcare.'

'Sure, like girls who get sick or steal your stuff or suck off their boyfriend in the middle of the day while your kid's locked in another room?'

'I don't want my son looked after by a bot.'

'So Frankdaddy has to play nanny while you're flying round putting Christmas and your whole life in danger, even though you could be working from home like a normal person. Huh,' she flops back in the seat. 'I'm starting to think old Frankdaddy's some kinda hero. How's he feel about this?'

'He wants an iMom. He wants reliability, predictability, and 24/7.'

'Jeez, if I could turn the clock back I'd sure rather have a bot than any real mom I ever met. And iMom kids are *way* smarter than humancare kids.'

'Yeah, well, Gurl, there are problems with that,' Scarlett taps the wheel. 'For one thing, iMom kids end up preferring the company of machines.'

'Good thing too, cos a human is just about the most unreliable creature on this planet.'

'And Fintan wants an iMom for Christmas. It's all he talks about, like a fucking obsession.' She elbow-thumps the door. 'And Frank is totally on board.'

'So what you gonna do?'

'He's getting a bike.'

'Hah,' Gurl flings her head back and slaps her thighs. 'You crack me up!'

'It's all wrapped, in the basement,' Scarlett snarks through gritted teeth.

'Oh, Scarlett,' Gurl twists round in the seat, pushes back her hair. 'You gotta get him the absolute top-of-the-range iMom – look just like you, same voice, same memories – '

'Leave it,' Scarlett's hand shoots up rigid as a sentinel in the space between them.

'Just saying,' Gurl mutters. 'You wanna hear some of the human nanny stories going round.'

'LEAVE IT!'

There were plenty of scare stories in the early days. The worst case, Scarlett recalls, was the iMom who called emergency services when her four-year-old charge sneaked out the back door to the swimming pool. The i320 was an indoor-only model and therefore powerless to help, since her robot body could not navigate water; nor, indeed, the steps that led to the pool. But it was able to livestream the event so the real mother could watch her little girl thrash, scream, and finally sink to the bottom of the pool. All iMoms come with the standard adult-supervision-only disclaimers, and in that tragic case it was Dad who'd left the back door unlocked in his dash to an appointment. The truth is, of course, that iMom malfunction is extremely rare – human error or negligence was, and is still, almost always the problem: failure to update software, failure to maintain, clone models bought for cheap on the black market. TechNeglect orders are frequently exercised and violation means fines, or worse – kids can be removed from the home. But it is not the scare stories that trouble Scarlett, it's the recurring nightmare that infects her dreams, the one she cannot purge from memory.

She had dozed off on the flight and yet again the hideous film snippet found her in sleep. *Well, what do you mean by saying a baby loves its mother? Let me show you a monkey raised on a nursing wire mother.* Harry Harlow's 1950s experiment in attachment theory: baby monkey106 was weaned on a wire mother but clings to the soft cloth mother for comfort, seventeen hours a day. Trembling and alone in his cage, he shrieks at the diabolical clanking metal monster, while Harlow in his white lab coat ponderously documents his reign of terror. *He's scared alright* – his voice laboured and slow, as if he was medicated or speaking to an idiot. Scarlett cries out, gulps awake, hears the swish of the attendant who has picked up on her distress. I'm fine, she tells the gleaming eyes that scan her biostatus. Just a dream, she forces a smile. A little something to help? No thanks, she waves away the offer of iRescue and tries deep breathing. Thinks of Fintan, but finds no comfort there, pursued by the imagined whisper of her simulated voice and the image of him cuddling an iMom, his little arms clinging to the avatar neck. She turns quickly sideways and lies watching the sleeping outline of another passenger across the aisle and copies his steady breathing: in and out and in and out, soothed by the faint scent of lavender and the gentle orange glow of the pinprick spots that illuminate her soundless flight.

'I'm sorry for snapping at you, Gurl.' Scarlett turns to Gurl, who sits sullenly studying her fingernails. 'So what's the problem with Blane? Sounded like he was angry.'

'He's just pissed cos I left Roxanne on her own in the house.'

'Your friend – she's not well?'

'She just needs help sometimes. With walking and stuff.'

'She lives with you?'

'Does now. Since a while back when she got dumped.' Gurl smooths her lacy frills. 'Roxanne's the best friend I ever had – always got a kind word,' she flips down the visor and begins a side-to-side inspection of her face. 'Like, back there, when I'm freezing my butt off and Blane is a no-show, even though he promised he'd pick me up. And I'm wondering how the hell

I'm ever getting home, when Roxanne sends me a sweet little message.' She picks up her phone and reads aloud: 'The only way to have a friend is to be one.'

'Emerson,' Scarlett smiles.

'What?'

'It's a quote from Emerson.'

'That's just what I'm saying,' Gurl looks up from her screen. 'Roxanne tries real hard to make you feel good. Here's another one: "Friends are born, not made." Ain't that sweet and true all in one?'

'You two must go way back?'

'Nope. Truth is, I hated her guts first time I met her. But that's a whole other story,' Gurl bends to slip off her boots. 'And right now I gotta warm up these ice blocks.' She peels off black socks to reveal cut-off leggings. 'You mind?'

Scarlett shakes her head and Gurl raises her right leg, stretches luxuriantly, and lays a small naked foot on the dash; slim and white with a silver ankle chain and coral toes. Bends forward in an elegant spinal curve and massages her ankle with both hands.

'Ever done ballet, Scarlett?'

'No. But looks like you did.'

'Started dancing when I was eight years old and I ain't never stopped. There was a ballet teacher lived next door those days. Rena Carter. Gave me my very first leotard that I still have – a real classy purple, like lavender. She was my mom's best friend, least for a while – hell, everyone was Mom's friend, till she went and lost them.' Gurl adds her left leg to the dash, holds them both straight as rods, and flexes the ankles. 'Rena used to teach me in her kitchen. We used the hanging rail for a bar. Oh, I learnt real quick – pirouette, pleeyay – used to twirl round on the linoleum till I got dizzy. Hon, you remind me of my own self, Rena'd say, standing there in her housecoat, smoking. Smiling at a picture of her twelve-year-old self on the wall in a yellow puffball and a tiara, arched on her points at the end of a show. You could just about recognise that little ballerina buried in the sag. Rena was real fat. Don't-care-fat I call it – you know the kind, Scarlett? When a person just stops trying. Back then all the washed-up women were skinny or fat. No one just in-between.'

Gurl sighs – a big pantomime sigh, puffing out her lips. 'Rena had this old wooden box tied with a silver ribbon and stuffed with straw – her very last pair of pink satin ballet shoes. Like your toes on solid rock. Course she didn't wear them no more. Couldn't arch her toes by then. Or even balance. She smelled of cigarettes and fried food. There was no man about and she was always sighing about that. You get yourself a good man, Gurl, she'd say, and fix my hair so tight it hurt. Her and my mom sitting there, swapping stories about all the men who left. Y'know, when I think of her back then, my mom was always nodding and sighing and pushing herself out of chairs and then sinking back into them. Like her body was made to be pulled down by some giant magnet. Her whole life rising and falling. Dragging herself round like a sack. Rena had a daughter same age as me. Myrna – jeez, what a dumpy sourpuss, never said a word, good for eating and nothing else. Two long stringy bangs hanging either side of her face. Titties sticking out even when she was eight years old. A slump like a hunchback. Big fat moon head plonked straight on those shoulders. Sitting there like a monster blub; hugging on a cushion while her mom taught me. And *one* and two and three and *hoooold*,' Gurl slices the air with her hand.

'Myrna's eyes be stabbing me from behind her cushion. Rena would start the music and off I go, being a swan or a princess. Dancing my way into being someone that was the real me. The *real* Gurl was up there in a spotlight twirling and spinning and holding still till my legs burn – till I think I can't hold no longer, but I learnt to push past the pain. The *real* Gurl was shiny, with all her hair in place. Dances like an angel, said Rena, squeezing her hands. My mom nodding her head like she had something to do with it. She said I was so good she could cry. Which woulda been just about the only time my mom ever cried over something that wasn't a man.

'Sometimes I'd dance so hard I'd get real tired and Rena'd give me a spoonful of honey. So sweet it hurts your teeth. Coats your mouth in a dry waxy sheen and makes you all sleepy. I can still taste the way it catches the back of your throat; leaves your mouth stinging.' Gurl runs a tongue over her lips, leans forward, and curls the toes on her right foot.

'What about Myrna's dad?' Scarlett prompts. 'Was he around?'

'Now you're asking,' Gurl shakes her head in slow wide sweeps. 'Meanest streak of badness you ever could meet, and Myrna knew well to shut her mouth when daddy showed. Which, lucky for her, was close to never and about twice a year. He'd pitch up late at night outta the blue. Him and Rena'd fuck and party for a couple of days before trouble started. Yelling and screaming, fighting about things, mostly themselves. He was fast with his backhand and his mouth and all the shit in his head. One time in the kitchen Rena was out at work and he took and yanked Myrna's hair so hard you could see her scalp lifting, bent her neck right back till I think it's gonna snap. She's staring, clinging on to the table making this godawful *mmmmnnaaaah* sound, and all the while he's smiling the biggest maddest smile you ever saw. His eyes are fit to pop and Myrna's blubbing and shaking and then he turns his head – he's looking right at me. Don't even try to pretend not to be ripping off my dress with his eyes. No matter that I was ten years old. Thu,' Gurl clucks.

'All round that neighbourhood was women everywhere you looked. And not a good man in sight. A whole stream of men pulling in or pulling out like we were a giant drive-thru. Women and kids all living there just waiting to get fucked up and over. Moms falling like skittles for any passing dick. Huh, mothers. Don't get me started. My mom used to talk Bible sometimes when she was drinking with Rena. God made a man then took his rib and made a woman. Gave her the special power of making babies, so the future depends on her. So you think men woulda treated their women a whole lot better. But God was not smart, Scarlett, because there is no power without muscle. That's your design problem right there. I guess God was just busy inventing a whole human race so maybe he didn't care what men did to women, long as they fucked.'

Gurl makes a spitting sound, twitches her head against the seat. 'Rena told me she had a stage name for me. Had it all worked out they said, her and my mom. Do I even need to tell you? Bullshit. Trash dreams. All they were good for was talking and eating. I swear. That's what I remember – eating, talking, and waiting for whatever loser guy had left them to come back. Which mostly they didn't. They just carried on sitting there waiting for some other asshole to show up. What I learnt early on: women are real good at waiting.'

'You tell a good story,' Scarlett nods at the windscreen.

'Oh, everyone *loves* a good story.' Gurl shakes her hair, tucks the floater wisps back into their bind. 'I got lotsa practice cutting hair down the salon. And me and Roxanne are *always* trading stories. Though it's mostly me, Roxanne never gets tired of listening.'

'So what happened to Rena?'

'Last time I saw her, she was crying in the kitchen,' Gurl nods solemnly. 'Standing there in her black leotard with her housecoat hanging open so you could see the fat rolls under her arms. All of a sudden she's leaving; all packed up in a hurry account of being kicked out for no rental. She's running in and out of the house, yelling at Myrna and trying to get her stuff out 'fore they come and take it away. So I sneak over where the trunk is open and I see the box with her ballet shoes. Take them out for one last look. And then I stuff them under my shirt and close the box so she won't notice.'

'You took her ballet shoes?' Scarlett turns open mouthed.

'Sure did. I wished so hard for those pumps and Rena sure as hell didn't need them no more. Just looking at them made her cry. Figure I saved her from future pain. From all the hurt that just kept raining down on her. All that dreaming and wanting.'

'But you took the one thing she loved!'

Gurl wipes a backhand over her side window. 'Scarlett, let me tell you something I know from the inside. Sometimes the thing you love is the problem. And sometimes the thing you love the most is the one you need to lose. Rena Carter is free now. And I like to think I set her free.'

'By stealing her ballet shoes?'

'If you got nothing, you can't feel the pain of losing. Was the past that was hurting Rena, all that failing piling up on her with the pounds. I did her a favour when I stole her past away.'

'You gonna answer that?' Gurl points at the vibrating phone. Scarlett glances at the flashing demand.

Y/N?

She presses her fingertip to silence Colin. Let him wait, she is not due to call yet. But the flashing returns and she fingerprints again. Pictures him simmering in Code; nipping on his thumbnail, scowling at the screen.

'Now there's someone wants you real bad,' says Gurl.

'How do you know it's a he?'

'Figures,' she shrugs. 'Yes–no, bullet questions. Some guy who don't know you too well, I'm thinking.'

'Actually, we've worked together for thirteen years.'

'Well, he sure as hell don't know how to handle Miss Scarlett Buick La Salle,' she grins. 'I'll give it thirty seconds before he's back again.' She leans forward chin in her wrist, staring at the phone. Scarlett knows Gurl is right, Colin will be back, of course, and yet it has always been curiously pleasurable to taunt him in spite of his predictable responses.

'What's he want so bad?'

'An answer.'

'Jeez, give me something here, Scarlett – he asking you to marry him?'

'A yes or no to a deal. Before I get on the plane.'

'So what's your answer?'

'Need to know.' Scarlett smiles, places a finger to her lips.

'Cos Gurl here is gonna tell the networks.'

A rush of birdwing skims the glass, divebombs, and disappears. Scarlett's foot tenses. Her eyes track its wheeling away towards the skies. Grey, insipid, limpid – she tries to think of a word that could refresh the jaded familiar and shift perspective on the monotony overhead. Graphite is the word that rises to the surface. *Graphite*, she mouths and repeats – sleek, smooth, seductive; a sensory pleasure to articulate. The Buick glides on past the brooding forest, though Scarlett thinks she detects a faint softening of the light that lends a charcoal sheen to the dark tree wall.

'Sure must be one helluva deal that puts you out here in Nowheresville, playing fast and loose with Father Christmas,' says Gurl, still monitoring the phone. 'Guess you're one of those ubertechies.'

'An accidental techie.'

'You really need all those secret caves in the mountains?'

'No choice these days, with all the cybercrime,' says Scarlett. 'Code is like gold dust. People stealing things you can't see. And people. All the important work is done in places like this. Secure sites, safe spaces.'

'I heard about that underground cave in the desert where they kidnapped those coders. Burnt their fingertips off. Eyuck!'

'The human enhancement lab.'

'You ain't worried about your hands?' Gurl wiggles her fingers. 'Whehey, Scarlett! Here's your man, right on time. I'll tap it for you – ' She reaches for the shivering phone.

'NO!' Scarlett grabs her wrist. 'He'll know it's not me and then he will really get paranoid.'

'How's he gonna know?'

'It detects any fingerprint that isn't mine and triggers an alert.'

'And here's me thinking your phone was some old piece of junk with all that ugly rubber.'

'That's the whole idea. Make smart things look cheap.'

'What's your man so uptight about?'

'Algosnatchers who steal code to sell or to snoop. Or to eliminate the competition. And the cyberactivists who call themselves freedom fighters, who destroy code to make a statement. They call themselves vigilante philosophers, but it really boils down to the old story: kill the bots before they kill you.'

'You don't think they got a point?' Gurl frowns.

'Years ago it was animal rights people breaking into research labs, setting mice and dogs free. Protest groups, pressure groups, death threats, bomb threats – this is just a different version of the same thing. People have always been trying to kill scientists or stall progress. It won't change things, just scares people. And pushes up the cost of doing business.'

'So that guy calling is your boss?'

'I haven't got a boss.'

'I'm likin' it!' Gurl reaches for a high five.

'We are three partners. That guy calling me is Colin. The other one is Felix.'

'Whoa, Scarlett – two men! So how d'you get this job?'

'I used to be a banker. Felix was my client. Colin was a techie. He worked at the same bank.'

Gurl flops back in the seat, fans the skirt of her dress across her thighs. 'Tell me the story of the little girl who grew up to be Miss Scarlett Buick La Salle.'

'Why are you interested?'

'Cos I like a good story!'

'There's no story, not really.'

'You know, Scarlett,' Gurl offers her a winning smile, 'one day you're gonna have to teach lil Fintan about supermom and the algosnatching. And I am interested! I know you ain't got all those rings through sleeping around, so why don't you just try tell me how you got from A to B without thinking I'm a retard.'

Scarlett unclenches her hand from the wheel and stretches. 'Like I said, there's not much to tell. I was a banker straight out of university. Worked on a trading floor for years. I was also a math whizz, though I wasn't really interested in that back then. Felix ran a hedge fund out East; he was the client who would only talk to me. I used to sell him bonds, though mostly he wanted to talk philosophy. Colin was the fat, brilliant nerd who used to follow me round the trading floor. Both of them were interested in my math thing for different reasons.'

'Got yourself a real circus there,' Gurl grins. 'So where's Frank, then?'

'He came later.'

'Was there ever anyone before?'

Will there ever be anyone after? Scarlett starts, ambushed by the silent question suddenly flung from the forest wall by some invisible mischief maker.

'So there *was* somebody before Frank?' Gurl seizes on the pause.

'There was a man I loved once. Who broke my heart.' She picks her way carefully over the words like hot stones, but even this simple utterance still burns. 'Though he was always the wrong man and everyone else could see that but me.'

'Love is blind,' Gurl murmurs.

'That may be a necessary condition.'

'What did he do to break your heart, Scarlett?'

Scarlett blinks fiercely at the road. Imagines the years unfurled behind her like a streamer. Will there ever be a moment when humiliation loses its sting? 'I don't talk about the past. It's a waste of time.'

Gurl nods, slow and hesitant like she's turning it over.

'Anyway,' Scarlett straightens up and pats the wheel. 'I was twenty-nine by then and I'd made good money. I'd had enough of banking and Felix and the fat man. Had enough of all men, maybe. So I went out West and spent a year in a university library. Blank slate, fresh start – all the clichés, since I had become one anyway. Felix found me a year later, just as I was starting to get restless. And then, one day, eight years ago, I bumped into this young guy on campus.'

She smiles at the memory of FlyBoy crouched over the white box on the grass. 'He had a great idea. And like magic,' she clicks her fingers, 'I found our project.'

Gurl opens her mouth but lets it close. Scarlett sees she is a gentle and attentive listener, flexing her way herself around the story like a hungry probe. Just like Fintan, she hangs on each coming word.

'So tell me,' Gurl nudges, 'why take the risk of coming out here just before your little boy's Christmas with all the NoFly problems?'

'I already told you: the deal. We have built something very special.'

'You and Felix and the fat man.'

'He is not fat anymore.'

'How'd he get thin?'

'He stopped needing to be fat.'

Gurl runs her hands over her thighs, smooths the ruched dress. 'So whyn't you just give your partners your answer and get it done?'

'I'm still thinking. *Over*thinking, I am told.' Scarlett stares at the long chain of freeway. The Buick gobbles up the blacktop, but it feels like the car is static and the dark forest flying past.

'No wonder Frankdaddy is pissed. Maybe you just gotta learn to learn to let go!'

'Funnily enough, that's exactly what he said!'

'Me and Frankdaddy'd get on like a house on fire!'

'He says I can't win, because it's two partners against one.'

'Then it's outta your hands. So just call your boys and give them what they wanna hear. Or maybe that's what you like to do with your men – hold out till the last minute?'

Scarlett glances down at Gurl's playful prod, the beaded band of turquoise and metal, the slim wrist.

'You gonna fall out with your partner boys over this, Scarlett?'

'I don't think so. There are other projects out there that we could work on.' She gestures vaguely at the windscreen, but the truth is she cannot conceive of an ending. *To love a thing is to know and love its nature*: to fall in love with your own creation. Volo's satin sheen soars and spins in her dreams, glitters against a black canvas. She sees it dipping and diving against the green walls of Test: FlyBoy strolling through his Eden, the lab hounds watching on the other side of the glass. *You need to let go*, she hears Frankisalwaysright echo in her ear. But Lab is her sanctuary, and where else in the world could she replicate the strange satisfaction of intimacy and remoteness? Felix and Colin are her homing beacons, her own personal gyroscope. And she's begun to understand that – much to her amazement – she may be at her best in virtual relationships. Work is the safe space where no one can break and enter her encrypted heart. The paradise playground where she finds her true self; takes refuge from human complications. *All work is an act of philosophy* – another favourite quote, though she keeps Rand to herself since no seems to care for old voices; there are new champions in this age of machine enlightenment.

'You gonna fall out with Frankdaddy over the iMom?' Gurl prods. 'Cos I sure as hell don't see you at home playing soldiers with Fintan.'

'You think you own this decision?' Frankisalwaysright had said, the night before she left, arms folded in the basement, watching her wrestle with the crackling reams of shiny red and silver paper, trying to wrap the unwrappable bike. 'You think it's yours to decide because you're the mother?'

Now that's a road she will not take, not right now just before boarding the flight. But 'yes' is what she instinctively felt. YES! That's the truth – and anyway it's such a ridiculous question. She carried Fintan, it is her decision and there is something darkly, solidly primitive – a zero-doubt heat in her gut.

'What I'm saying,' says Frankisalwaysright, 'is that it's totally inappropriate for Fintan to bond with a woman who is not his mother – a human nanny, someone with whom he will have NO relationship whatsoever in the future. Nannies pass through children's lives and disappear – boom! So tell me, how is that good for a child? You really want Fintan to attach to a woman who is paid to look after him, has no vested interest in his life or future, and can fuck off to another job whenever she likes? Far, far better and healthier for him to have an iMom avatar of his real mother, who can stand in while you are away.

Who talks like you, tells stories like you, has access to your memory bank – '

'Who is a fucking machine,' she yells. 'Who is NOT ME!'

Frankisalwaysright steps close, takes her hands in a gentle grip.

'You're not the one who comforts him at 3 a.m. when he's crying his heart out. You're not the one who sees him in distress, for no good reason. Look, I get that work is central to your life and I'm 100 per cent behind you, always – you know that! But a human nanny is unreliable, unpredictable, and incapable of delivering the best care. Why trust in human frailty when we can have an iMom who is totally dedicated to Fintan?'

He squeezed her arms as if to elicit response. She glared at him; he read it and his hands slipped away. 'Well, I'm done with this,' he pulled back. 'Love and work, work and love, you're always saying. If you want to keep on travelling when most mothers don't, we give up on the human nanny and get the iMom. So, you get to do what you want, you just do it a different way.'

'You going to fall out with Frankdaddy about the iMom?' Gurl repeats, reeling her back into the now.

'I've already fallen out with my best friend.'

And Gurl nods expectantly, waiting for a story, but Scarlett will not speak – three weeks old but it is still too raw. She looks down at her white-knuckled grip, her hands ache from over-clutching the wheel. The road ahead, the road behind, she flexes her fingers, glances at the mute phone that awaits her response; her countdown to the end of things.

'Scarlett,' Gurl touches her driving arm. 'You thinking maybe Frankdaddy's gone and got the iMom already and dumped that bike?'

'He would not do that,' but Scarlett hears the tremor in her voice. Imagines Frankisalwaysright standing on the other side of a deep crevasse, a frayed rope bridge in between them. Cut or compromise. Recalls the basement gloom, her toes curled on the chill wood while he was backlit and inscrutable; and in the morning she was gone.

Gurl circles her finger on the dash, tracing spirals ever expanding. 'Tell me about that man who broke your heart.'

'I'd far rather listen to your stories.'

'I already told you 'bout Rena Carter. I'm an open book here – so your turn now, Miss La Salle.'

'I've got nothing interesting to tell.'

'You got a life, you got a story. What you mean is you don't trust *Gurl* with your story. Look at you sitting there, tight as a clam. Gripping the wheel when you could be chatting. You gotta give me something.'

'I'm giving you a lift!' says Scarlett.

'You got a big secret?' Gurl prods her arm again. 'Something you ashamed of? Or you just too snooty? Like your story is worth so much more than mine. Like yours is a classy wine and mine just spills out like cheap beer.'

'I don't have a neat life story,' Scarlett waves a dismissive hand. 'And I already told you, I don't see the point of going back.'

Gurl rearranges herself in the white throne, leans back against the door, draws up her knees, and faces Scarlett side on. 'Tell me something you never told anyone.'

'You mean a secret?'

'A secret never told.'

'Why?'

'Oh, quit stalling, Scarlett!'

Scarlett checks the rear view. The gauge. The side view. The iron sky streaked with silver. Nothing behind or up front on this road that is in perfect symmetry with Now. A moment without past or future.

'Don't matter what you say, here on this trip.' Gurl's face is prickly flushed, her eyes gleaming, arms wrapped tight around her knees. 'This, here, right now, is your one-night stand. So c'mon, Scarlett. Tell me your story,' Gurl whines, needling, like Fintan's voice. *Mummy, another story.* 'Take a risk for once in your life.'

'I take risk all the time in my job.'

'That's work risk.' She slithers out of the puffer and flings it in the back. 'Real risk is *human* risk. Living in the moment. Getting down and dirty.'

Scarlett nods. 'You're right about that.'

'Here, I'll make it easy for you.' Gurl removes the scrunchie and loosens her pale hair and grasps it quickly and re-binds with a snap. 'Take me way back, before your tech days, before you were Mrs Frankdaddy, Fintanmommy, Miss Scarlett Uptight Buick La Salle. Roll back the years, just like I did in Rena Carter's kitchen.' She raises her hand and points an index finger at Scarlett's temple. 'You're eight years old. Tell me what you see.'

'I don't want to be eight.'

'Well there's your story!' Gurl claps her hands delightedly, 'The little girl who doesn't want to be eight!'

Scarlett pats the wheel. Gurl shifts, folds her legs into a lotus position on the seat. How flexible, Scarlett smiles, this ease of manoeuvre, to have limbs so biddable like those malleable dolls whose legs can be contorted into grotesquery.

'You know, I saw this thing once where you should write a letter to the kid you were,' Gurl continues. 'Make your own little self feel better in the past. So you can feel better about your grown-up self in the Now. Seems to me like you might wanna have a word with Scarlett aged eight.'

'I don't believe in all this stuff about the formative past.'

'What you want to forget about being eight?'

'Nothing,' she glances at Gurl. 'Sorry to disappoint, but there was no big drama.'

'Maybe you just don't like being told what to do.'

'That's true,' Scarlett grins.

'Ole Frankdaddy's sure got himself a real handful. But you know, Scarlett, after this ride you and me won't see each other ever again. Hell, we don't even know our real names. Here's your chance to open up when it don't have to matter. Have some fun. Every single moment in life is a chance to do different. I ain't never gonna see you again, so here's your chance to be *you*, before you were Mrs Frank and Fintanmommy.'

'Like little kids trading secrets.'

'Way more exciting when grown-ups do it.'

Scarlett feels a sudden lurch in the pit of her stomach, as if they had flown a bump in the road. Catches her own eyes in the mirror: her unrecognisable self. She frowns – at time, at speed, its passage-taking. What is this privacy she so jealously guards? No skeletons, just the simple fear of falling short of your own reflection. And what is disclosure, other than the fear of being known? Isn't that what we all crave more than anything? To be loved, validated, to fascinate, intrigue, seduce –

'There you go, thinking again. I swear you need an off-switch, Scarlett.'

The steering judders at some unseen change in the terrain. Scarlett tightens her grip, frowns at the freeway sucking them up, empty as the dull sky. She purses her lips sourly. *Blank slate. Fresh start and all the clichés. Since I had become one anyway.*

Oh, how neatly we package our past! The glib storyboarding that is the stuff of fiction, the jaded, repeating plots. The half-truths we make of the mess that actually happened. What *do* you learn from a broken heart? How much it hurts, is all. The past submerged like a shipwreck, you know it's there, but diving can give you the bends. And Gurl is right, of course: the moment is all we have. Here, now, with a stranger in a golden car, slicing into a future unknown. The real survival skill is the art of living in the present, in a state of continual renewal. The past schlepps ten paces behind; history cuffed to your ankle, threatening to hobble your next step.

Gurl's fingers prowl along the console. She is restless as a toddler, tapping the sculpted curves as if she might discover a secret panel.

'Mini-meds and ice gloves,' she rustles impatiently in the glovebox, 'strawberry protein bars – yuck, lemon is *so* much better – electrolytes, and uh-oh, Road Kill Kit.' She tuts, shaking her head, 'now *that* should be in the trunk. It's the law round here you know.' She slumps back in the seat, her hands flutter over the jumble on her lap like discarded toys.

'What are you looking for?'

'Oh, nothing, everything. Sometimes it's just nice to be surprised.' She twitches her nose; sideways on it is sharp, the skin stretched tight and shiny over the bony ridge. 'Whyn't you at least do Cruise?' Gurl frowns at Scarlett's driving hands. 'Let go for a change.'

'Don't like Cruise.'

'Or you just don't like losing control.'

'You said. That's getting old.'

'Whatever.' Gurl gathers up the glovebox jumble and stuffs it back in, slams the lid. Fidgets moodily with the beaded bracelet, twists and presses hard into her flesh then releases to inspect the red marks. The restless child who bores easily and goes looking for trouble. She takes a small silver phial from her tote and sprays it on her wrist, sniffs loudly: 'Mhmm, amber and figs, how 'bout that. Smells musky, fleshy, like afternoon sex.' She wafts her wrist in front of Scarlett's nose. And Scarlett takes this as the taunt intends, an arbitrary assignation of roles by innuendo. Scarlett, the dull moneybags brain of enterprise, and Gurl, the funchild – careless, casual, oozing suggestiveness and sex. And with a crackle of unpredictability: a flame that burns bright and dangerous, the good-time girl-woman you wish you were.

'Bet you know fuck all about surprises.'

It's a mumbled bait from the passenger seat; a provocation that Scarlett decides to overlook.

'So, tell me then, Gurl, what would be a good surprise for you?'

'Five thousand cash,' she answers, fast and quick like it's the question she's prepped for. 'In hundred bills, so it looks fat and warm.'

'And what would you do with all that cash?'

'Strip off all my clothes and lay those notes right over my bare body.'

The road inclines; the Buick glides magnificently upwards. Stretched right across the windscreen, Scarlett sees Gurl prone naked on a bed, whitely gleaming.

'Once a man did that for me already.' Gurl turns to face her side-on.

'Gave you five grand in cash?'

'Course, now you're thinking what'd I do for that.'

'That's exactly what I'm thinking.'

'Wasn't fucking. Though we *did* fuck – just not that time.'

'Then what was the money for?'

Gurl tilts her head back against the white leather throne and closes her eyes to black out the Now. 'He's sitting right there in the armchair in the corner. White shirt, loose tie, that golden hair. I'd like to be *bestowed*, I say to him. I'd heard that word. Real elegant and I liked the sound of giving for its own sake. I'd like, just for once, to get money for nothing, I tell him. Money I never earnt.

'He leaves the room, he's gone a half hour, and then he comes back and puts it on the table. I count – each bill is new and stiff and smells inky. I take off all my clothes and lie on the bed. He starts at my feet and works all the way up, real slow, covers me butt naked in money. But he knows to keep the moment pure. He doesn't touch me otherwise.'

Scarlett glances at Gurl's bowed head, sees the still life: Gurl's yellow hair fanned on the white sheets. Her pale flesh beneath a quilt of notes.

'His name was Nicholas.'

'Where is he now?'

'He's gone,' Gurl sighs into the window. 'And now there's Blane.' And the name hangs there between them, a swelling presence in the car. What's coming, like the phonics in Blane – a long narrowing, like going blind into a tunnel.

'FUCK!' Scarlett yells, swerves to the left as the bird thwacks the screen. 'Jesus,' she rights the wheel.

'Crows,' says Gurl, unmoved. Nods at the greasy smear on the glass and a black feather on the wipers. 'They've been dying round here lately. Get dizzy all of a sudden and fall out of the sky like stones. No one knows why. Julia says it's an omen.'

'Who's Julia?'

'She's my psychic.'

'You mean someone you pay to tell you bullshit.'

'Yeah, well, I might have known you'd say that, Scarlett. Course, you're so clever you can figure it all out for yourself. So you got the whole future nailed?'

'No. But I *do* know that Julia is taking your money just to make you feel better.'

'And you're so smart you can't even answer yes or no to a deal.'

'Maybe I should ask your psychic.'

'You got no imagination,' Gurl shakes her head. 'Or maybe it's all tucked away in your cold heart. Where do keep your dreams Scarlett – in cryo?'

Gurl flops back in the seat. Scans the car as if she is reappraising the interior. Glares out at the grey. 'Why don't you hit the override and change the view to sunshine? Brighten things up a little in here.'

'Nope.' Scarlett feels like being churlish, although refusal will only fan the incendiary restlessness in the air.

'You got way too much respect for reality.'

'That's because *this* is how it is,' Scarlett slaps the wheel. 'And *this* is where we are!' She jabs the windscreen. 'This is the world we inhabit, this is what we have.'

'Don't mean we got to keep staring it in the face,' snaps Gurl. 'Sometimes bullshit is exactly what you need to see and hear. Julia could be as right as anyone on there,' she jerks her head at the disabled infotainment system. 'All that matters is what you believe, right?'

No, that's not right, not right at all, Scarlett wants to say, but instead she rolls her shoulders. Twists at the waist. Leans forward to the wheel, trying to stretch her spine. Too much sitting, too long, and more to come on the overnight flight. She will have to remember to stretch off at the airport; can already feel the dull ache of the plane seat.

'So what's your psychic's advice, anyway?' she turns to Gurl.

'Gave her up a while back. These days I just ask Roxanne what she thinks. She knows *so* much stuff.' Gurl picks out her phone. 'Now I got to call Blane and tell him how far we got before he starts getting all antsy.'

'You won't get a call signal here. Text will work.'

'Blane is not a man who likes to read.'

'My phone has a special booster.' Scarlett nods at the dash. 'You're welcome to use it.'

'He'll go crazy if I call from another number.'

'Doesn't trust you?'

'He's a guy,' Gurl shrugs. 'You telling me Frank trusts you? All this travelling, you could be doing anything.'

'Except I'm not.'

'You ever had nothing, Scarlett?' Gurl's voice is newly thin and jagged. 'No, I don't suppose you have. Bet you got plenty to lose, though.'

Scarlett frowns. 'Where are you going with this?'

'Bet you could cover yourself in your own notes any time you want, Miss Scarlett Buick La Salle.' Who registers the sudden tone change as an alert to some failure in transmission; a fault in the code.

Scarlett turns her head to meet Gurl's steeling, lips pinched tight against her teeth. Eyes aglitter, as if a torch flame burns inside her skull. 'Bet you can't even imagine a person doing something for nothing,' Gurl spits. 'Bet you think it always has to be a trade for a trade.'

'So you're telling me that guy Nicholas wasn't getting something out of covering your naked body with his own notes? That it had absolutely nothing to do with money and sex?'

'What do you know, Scarlett? You think there couldn't be a guy who would just do that? Just *bestow* for someone else? That just *maybe*, Nicholas giving me that money was an act of grace.'

'Oh, sure,' Scarlett snorts. 'It had nothing to do with you getting naked.'

'So money makes it dirty? Well, *you* sure as hell got plenty. Bet you even think I want what you have, like this stupid car that can't even drive itself,' she punches the plush leather. 'And that big showy diamond ring on your finger that needs its own freaking transportation. And your high-end haircut that's the wrong one for your face anyways. And you being tired and scrawny and looking years older than you probly are. And your lil Fintan – '

'What the fuck is your problem?' Scarlett whips round. Gurl's face is weasel-sharp, lips parted to show small white teeth, throat flushed red.

'And you going round in this big golden disconnected the day before Christmas and your little Fintan missing you like hell, though you won't even let him have an iMom. And old Frankdaddy wishing you was home.' Her voice shrill and loud and darting like a crazed hornet unleashed in the Buick, 'And your big job and your great big sparkly life on the other side of the planet. You think money is just for buying and owning. So when was the last time you ever did something for nothing?'

'Why don't you just shut it.' Scarlett's heart thumps against the impulse to lash out. Exhales slowly to steady herself against the snarl within.

'And FYI, I know why you don't want an iMom,' Gurl smirks. 'You're just afraid Fintan will love her more than the real thing.'

Scarlett blinks, squeezes her eyes tight, squints at the road ahead. What is this ludicrous tear unrolling to her lips!

'Oh Lordy,' Gurl says, disgusted. 'Now she's gonna blub.'

'SHUT THE FUCK UP!' Scarlett whacks the wheel, 'and get your fucking feet off my dash.'

Gurl grins evilly, folds both legs beneath her, slick as a shape-shifter. Tucks in her earbuds and checks out, closing her eyes.

Scarlett grips the wheel to stabilise, searches for solace in the world outside: the unrelenting cloud, the black tree wall. Lets her eye imagine the gift of the familiar and the invisible.

Those natural mysteries that lie hidden in the forest floor or the seabed; where creatures thrive or disappear, leaving their clues. REPORT ROAD KILL – BY LAW, the sign approaches on her left, a red-flashing cross hairs with a silver silhouette of a generic creature. She recalls a rare trip with Colin last year; how they screamed to a halt when a bobcat blundered towards her tyre but collapsed before impact. They stood watching on the side of the road while Colin filmed its twitching expiry and sent through the report.

'Though it's a joke,' he sniffed. 'These reports don't even get processed. The guys at Z4 are working on this and they are choked with data.'

She snapped on the rubber gloves and hunched over to study the corpse, so emaciated its skeletal structure was clearly visible. Bald patches of skin all over the body, except on the forelegs, where the black and grey stripes were curiously intact. A clump of fur whisked away like tumbleweed by the wind.

'Chytridiomycosis,' she murmured. If they hadn't found Volo, this would most likely have been their project – the chlamydial zoonosis laying waste to the wildlife, that started with the vanishing salamanders and was busy crossing species a decade before anyone figured it out.

'The biggest threat to global biosecurity ever known,' said Colin peering over her shoulder.

She stared at the bobcat's dull yellow eyes, tried to imagine him sleekly whiskered at the moment before he began to fail. The first sign would be anorexia, and then a lethal lethargy that would have him labouring in slow motion and disorientation, sitting in odd positions, failing to seek shelter, failing to flee. Finally, his fur would drift away and the naked skin thicken so he could no longer take in nutrients. Or even breathe.

'Death by choking,' she stood up.

'It seems the Buddhists are to blame,' Felix had told her earlier that week. 'According to some fabulously bleak data I have just seen that is not in the public domain. Are you familiar with the practice of animal release?'

'Tell me.'

'Fang Sheng: be kind and compassionate to animals. Liberate the caged birds and the farmed fish and the pet amphibians and return them all to their natural environment. Where, of course, many of them will perish. There is a ceremony where the monks grant the creatures two wishes. One, that they will not be recaptured. Two, that they will be reborn as humans in their next life. In which case, they would be perfectly qualified for employment in the lucrative business that led to their demise: the deliberate capture of wild animals for release.'

Scarlett turns to Gurl, curled now in the sumptuous leather cushioning, with eyes tight shut: squeezed in an effort to make her feigned sleep convincing. How young she looks, just like a child. Scarlett smiles, restored. She skims a slow palm over the pale wheelskin, imagines the satisfied purr of the precision engineering that propels them forwards. It is this ability to distract through digression that always nudges her towards the recovery position. A habit, long practised, that filled up her scorecard of success – the sound of hands clapping, the CV swelling line by line. Tick, tick. First to the post, first past the post, summa cum, etc. Though what looks like a progression onwards and upwards has taken some very wrong turns – like this very moment here on the freeway, driving some stranger to a place she doesn't know, undone by the prick of tears. How pathetic, she thinks; how lame. The kind of stress response that

raises an alarm in biofeed. Volo, endings, her own intransi-
gence, blah, blah.

And then it hits her. She reaches out to poke Gurl's
shoulder, who unpops her earbud, rolls it between finger
and thumb.

'What?' she snaps.

'I am doing it.' Scarlett smiles. 'Right now. Right here.'

'Doing what?'

'What you said: I am doing something for nothing. I am
driving a person I do not know to a place I have never been.
Just because.'

Gurl's lips spread broad and wide. 'You surely are. And
there's my surprise!' She laughs, claps her hands gleeful as a
child, leans over, and kisses Scarlett on the cheek.

'Oops, steady,' Scarlett leans into the wheel.

'Oops,' Gurl mimics. 'You kill me with that cute accent,' and
she settles back in her leather throne. 'Nicholas used to talk
like that.'

'So he was from – '

'Yup.'

'You never said.'

'Wasn't important to the story.'

But it changes it somehow, in a way that Scarlett can't say.
She imagines Nicholas in a crisp banker's shirt, a buttoning
that invites the man to be undone; something beautiful in
stripping away and undressing the body. Memory flashes and
whips back six years to the first time with Frank. Room 2108, a
steely eastern dawn in the full-length window. A puce cherry in
a cocktail glass that she could see from where she lay, listening
to the sigh of an elevator shaft, or perhaps it was a tremor. But
seismic shivers had become routine. Frank slept and so had

she. A small miracle in a life stalked by sleeplessness and the first time she had ever drifted off in a man's embrace.

'You are a narcotic,' she said, and Frank laughed. Maybe she was thinking in cliché, but there really was a twinkle in those blue eyes that would become Fintan's gift. 'I mean, I can actually sleep with you. I've never done that with any other man.'

'So I win the sleeping prize! I guess that means we're forever,' he said, an untroubled blue.

'Now you're a clairvoyant?'

'The future is made, not guessed, babe. So let's get on with it.'

In their few free hours, they visited a temple, ate noodles, and basked in a sea of chatter they did not understand. They wandered into some kind of design gallery with small, roughly sculpted water cups that grazed the lip, sank into low leather chairs, and napped in the late afternoon. In the space between the bed and the bathroom mirror she had found herself transformed. Pale body aglow in the mirror, lovemarked by his touch, his smell seeped into her skin as if it had always belonged. And so it came to pass. So it was. And that was the beginning of all this.

'Where did that come from?' Scarlett stares at two blazing headlights crammed in the rear view. 'Right out of nowhere, I didn't even see it.'

Gurl peers into the rear cam. 'Yep, he's real close.'

Scarlett watches the black pickup in the side view where the headlights flare huge and round. 'Why doesn't he just pass me out?'

'Cos he ain't looking to overtake you.'

'What d'you mean?' Scarlett accelerates.

'Didn't they tell you to get blackout windows in your rental, so assholes can't see in?' says Gurl. 'Just keep driving, keep looking straight ahead. Any minute now he's gonna pull back, let you think he's losing interest and then – when I say NOW – you floor the gas. How fast this thing go, anyways?'

'Don't know.'

'You gotta put some distance. This here is a game you gotta win.'

'What game?'

'This is good old-fashioned pussy-hunting, Scarlett. What else you do for amusement if you're a coupla losers with a pickup? That's what guys do round here when they're not shooting up Road Kill signs for target practice.'

'Are they after me?'

'Hah!' Gurl snorts. 'These guys ain't looking for algos. This ain't no cybercrime. They looking to give us good old scare, then run us off the road and get busy. And if they're smart as well as nasty they send up a swarm of funbots to land on the screen and then you got no choice 'cept to stop.'

'Have they got – '

'Guns? Yep. This here is *sport*, Scarlett. You got a shooter?'

'Fucking hell, Gurl, of course I don't have a gun!'

'Just asking. You're the one going round with a cyber target on your butt.'

The pickup falls back as if it's rolling back down the slope. Gurl leans into the wing mirror. 'NOW,' she yells. 'HIT THE FLOOR.' And Scarlett floors: they shoot forwards, she watches the pickup fall away in the mirror.

'He's losing interest.' Scarlett hears the tremor in her own voice, watches the speed tick 90, 92, 98. The pickup is toy-sized.

'He's just playing, trust me. So get some distance. Don't be fooled, he's got a real big engine.'

'We could call the police?'

'Yeah, right, like "Hello, 911, my name is Gurl. Me and my girlfriend Scarlett are on the freeway in a big gold Buick La Salle and there's a coupla guys in a pickup trying to run us off the road. So we were wondering, like if you ain't real busy catching cybercriminals, could you send a car over and tell those bad boys to stop doing that?"' She puts down the mock phone. 'Cops say to give call back when the boys get us off road and they ready to get down to pussy business.'

They crest the hill and look down ahead at a long straight slope.

'No sign,' Scarlett straightens up, checks the dial. 'Maybe they did lose interest, got distracted by something.'

'Yeah, like a woodchuck.' Gurl mutters peering over her shoulder. 'OK, downhill and straight.'

Scarlett tightens her grip.

'Your guardian angel!' Gurl nods at the pulsing code, the siren rises to a piercing scream. Scarlett prints and it falls silent, returns immediately, and she prints again. Headlights glimmer like pinpricks in the mirror; effortlessly gaining ground, as if the pickup is on super fuel.

'You good under pressure, Scarlett? Cos here it comes.' Gurl leans into the wing view, belts up for the first time. 'Outside lane, NOW!' Scarlett swerves. 'Stay right here, and don't lose your nerve. This is a face-off. That pickup's heavy and these guys are nuts. Cars, pussy, whisky, I been here before.'

'And?'

'Still here, ain't I? Forget him and drive like you're in charge. Eyes in front, on the future. Don't look to the side, no matter what.'

The lights fill the rear view and a bullhorn roars deep and bass and loud, then a scream of heavy metal. The Buick shunts, Scarlett screams. 'White line,' yells Gurl, 'white line and nothing but!'

The headlights vanish.

Gurl turns to Scarlett. 'He's gonna disappear on your blind spot and pull up on my side. Real close. So don't look – just keep that wheel straight.'

And there in her peripheral, the closing shadow of approaching metal. The road sucking her speed, the trees flung past like a brooding army.

'Dead straight!' Gurl snaps savagely. Scarlett's hands are fused to the wheel, the judder and pull of steering just a whisper from the reservation, the line, the black blur on her right, a metallic rip, sparks flash. 'HOLD STRAIGHT. KEEP OFF THE RAIL!'

A gap opens on the right, the black shadow slips, slips, slips – and disappears. The sloping road levels flat.

'He's dropping back.' She watches the pickup fade in the rear view: 30, 40, 50 metres.

Gurl twists. 'He'll come again. One last shot.'

'I can't.'

'Yes, you can and you fucking will.'

He closes in, draws level. A window opens, a thick hairy arm comes out, he slaps his door and salutes a thumbs up. The pickup powers easily ahead.

'Gone,' Scarlett croaks, easing back on the gas.

'Sure is,' Gurl nods.

They have sunk to 22. Scarlett's hands quiver. 'I have to stop. I have to – '

'It's all over. You did real good,' Gurl pats her trembling driver's knee.

Scarlett shakes her head. 'Dizzy,' she gasps and veers right to the hard shoulder and stops, hunching over the wheel in a white-knuckled grip. And it seems as if the Buick sighs in relief. The ashen road, the steely sky, and the sour taste in her mouth. The phone flashes and wails.

'Just tell him you're OK,' says Gurl.

Scarlett thumbs the screen, but it flashes again.

TALK

She raises a silencing finger to Gurl and presses again. Swallows.

'Why are you stopped on the middle of the freeway?' Colin snaps.

'Just dizzy.'

'It's not safe – you need to move on.'

'I know, I know.'

'What's the problem?'

'Nothing. I just needed a moment.'

'Stop blanking me.'

'I answered the phone, didn't I?'

'So is it yes or – '

'Drop it, Colin, or I'll cut my chip again. I'll call you later, like I said.'

The screen blacks. 'I'm getting out.' She turns sideways to reach the door.

'No way,' Gurl grabs her arm. 'Open freeway is the last place you want to be. Plenty more where they came from. Here, let's put down the window – some chill wind to sharpen you up. BREATHE!' Scarlett takes a lungful. 'And slooooooow release,' Gurl models, blowing through her lips. 'Good, that's good,' she pats Scarlett on the back. 'You're gonna be fine.'

'How can you stick living in this fucking shithole.'

'Yeehawsville. All the privacy you techies ever wanted.' Gurl pulls Scarlett's right hand from the wheel. 'Stretch out those arms and wiggle.'

'Reminds me of those movies – you know the one?' Scarlett rolls her shoulders, swivels her neck. 'Two girls riding along in their car, all happy?'

''Cept it wouldn't be a Buick La Salle,' Gurl grins, 'it'd be an open-top number in the real actual summer – and the girls be singing along to the radio, wearing short skirts and looking all ditsy, showing lotsa leg, and fooling with their hair. Which'd be long and blonde and flapping like scarves in the wind, when two guys pull up longside in a pickup looking like wolves and maniacs and the girls start flirting and playing with their hair till the pickup starts bumping them off the road. And then – '

'They're screaming like pigs.'

'Till they hit the ditch. With cut lips and a few scratches but looking all sexy with frizzy hair and tops slipping off the shoulder like they undressed by a car crash.'

'And wolf maniac leans over and says?'

'You my little bitches, now.'

'Oh, pass the popcorn,' Scarlett laughs.

'We gotta move,' Gurl says.

'Look,' Scarlett holds out her hand, 'I'm still shaking.'

'Every minute we spend here just puts us in trouble's way. You got a little boy waiting for you, so dig deep. There's a place up ahead, a diner I hear is kinda retro, though I've never been inside. It's not too far, so get going.'

Scarlett squeezes her eyes tight and starts the engine. The Buick glides forwards. 'OK,' she nods at Gurl. 'But I need some distraction. I drive, you talk.'

'Oh, I got plenty of stories,' Gurl grins.

'Then tell me the story of how you met Nicholas.'

'Oh, sweet sixteen, there you see me

at the back of the class with Lester sitting right behind me and his hand snaking under my T-shirt. I'm trying hard not to squirm cos he's tickling right on my spine and I sort of like it and I sort of don't. Lester's a jerk but he's got real good hands. He did this last week in English and I got in trouble. But I'm always getting in trouble – the head says I go looking for it. He could be right. Can't keep my mouth shut, can't stop dancing in the corridor, doing the splits, showing it up to the teachers. Twitching my butt, making them want to look. Teachers hate you for that.' Gurl wraps her arms about her legs and her chin on the dancer's bony knees.

'I'm looking out the window where the leaves are turning and I'm flexing my calf even as Lester's hand creeps up my back – there's a new teacher starting a dance class after school. Not ballet, just general dance, but I signed up. Rena Carter's long gone but I'm still dancing, still dreaming.

'Lester tries to snap my bra strap and I'm wearing a real thin tee – if he does that I'll be in trouble – so I shuffle forwards, lean over my desk, out of reach. Lois is watching, smirking in the row beside me. Mr Haver, he's talking about muscular poems. My chair scrapes, Mr H stops and Lester's hand freezes on my back. I sit still as a rod and stare straight ahead. Mr H goes on and there's Lester's hand again, slow sliding down, like my skin's the most precious thing his touch has ever known. He's tracing each little nut on my spine, his fingertip is real gentle and soft, feels so good, and he slips lower, runs his finger along the elastic of my leggings, draws a line on my skin right round the side of my waist. I get that warm pussy feeling.' Gurl leans her head back against the window and rolls it slowly side to side.

'Oooh, I can still feel it right now, Scarlett. Lotta guys don't *ever* learn the power of soft hands on a girl's skin, and Lester gets to my side and I am still as post. I ain't moving, his touch is good and dreamy and Mr Haver's filling up the room with his poetry and passion and it's heating up and I know Lester has a boner for sure – he's *got* to have a boner now – and I'm thinking about that and trying not to. Lois is bug-eyed, lays her head sideways on her hands so she can watch. And I'm watching her watching Lester's hands, his finger sliding down real slow and I'm leaning so hard into the desk it cuts into my tummy, making a shelf for my ribs. I squeeze my ass towards him till he has that finger wedged real good between my cheeks. And I am nearly flat out, face down on the desk, leaning on my elbows.

'I hear Lester breathe, bend my head, and look through the crook of my arm so I can see his boner stiff in his sweatpants, and he's thrusting his hips – small little movements. And I'm clenching and squeezing and his finger is going lower, hooked into my panties, and he's pulling them down slow, so slow, I lean hard on my elbows so I can lift my ass from the chair and he can get where I need him to go. He got his whole hand there, his fingers spreading me, Lois's mouth is open and hungry for some of this; her tongue sliding over her wet lips, her shiny eye and the back-row boys all watching Lester's hand, got their own hands cupping their dicks, all of us in a sex cloud. I picture how it looks with my ass up and my panties pulled down, what they all can see, and there's breathing and grunting and dream lust in their faces.

'Lester squeezes my cheeks same time as I'm squeezing and slides his finger forward and he's there and I am rubbing myself good and slow, back and forth, working those dancer's hips and aaahh, I'm so wet now there's nothing going to stop me; then *just* as I come, Lester's hand is gone. I open my eyes, look up and there's Mr H standing right by my desk.'

Gurl twists her bracelet, stretches like she is testing its limits. Scarlett grips the wheel, holds her breath, poised on the hook. Pictures Mr Haver, gives him brown chinos, thinning hair, athletic poet-type looking down into Gurl's sixteen-year-old face.

'You want be some kind of porn star?' Gurl barks. Scarlett jumps. 'Headmaster's face is red then white then purple, and he can hardly sit in his chair. Is that the career path you got mapped out for yourself, young lady?

'Yeah, me and Lester, the double bill, I tell him. Never could shut up with the backchat.'

Gurl folds her arms tight and high, gripping her shoulders. 'Inside I'm dying, like a little bird. But I can see me now sitting in his chair all attitude, leg cocked up on my knee like a boy. Twirling my hair – it was real long then, and messed up – took hours to do good bed hair. Huh.' Gurl turns to Scarlett. 'You get a bit older and one day you look in the mirror and you see you've been going round for years looking like you just got fucked good and hard. Ripped tee over ripped lace, like the bugs ate right through it. Skimpy leggings you just about managed to pull over your ass. Yeah, the just-fucked look. Like that's the kind of life you want for yourself.

'Outside the headmaster's window the fourth graders were piling out for the bus; all tiny, singing and laughing, pushing and swinging their bags. I remembered that – it didn't feel so long ago to be so little. I swear I could feel the years rush up through the window and grab me by the throat. And all the while the headmaster is going young lady *this*, young lady *that*.

I wanted to say BUT. Has your body ever felt so *good*, have you ever felt the burn of wanting, just pure desire, I mean that's a human miracle, right? The fact that you can come?

'An act of beauty, is what I wanted to say. A gift, for which no person should have to apologise. And people watching don't have to make it dirty. 'Cept round here it does.' Gurl pushes back a strand of hair; her lips pinched tight and wistful.

'Indefinite suspension is what I got. Didn't think that was such a big deal till a week later when I got expelled cos I wouldn't apologise for the act. Everyone talking about it like it was something real cheap. And the guys? The guys think you *are* cheap. You become that one thing you did. All the fallout, all those things that happened after that I never coulda fore-seen. How you *become* the story. But it ain't the true you, it ain't the whole of you. You know, Blane moved from another place and Gurl was just about the first story he heard when he pitched up. Just how many good stories you think there are in Nowheresville?

'Even now, years later, his booze-hound pals look at me that way when they come round. I know what they're thinking when I walk past the screens. Sometimes I work it with my dancer hips, just to get 'em going. Course, then Blane has to lay claim. Shoves me into the bedroom. Tells me to make plenty noise so they can hear.' Gurl snorts, faint and mocking, stretches her arms out front. Scarlett tightens her lips on words that will only push her down a rabbit hole.

'Then there I am. Sixteen, kicked outta school. All the parents, all the dads, all the men thinking the same thing. Can't look me in the eye cos all they see is my ass and Lester's finger inside me. And they want to put their fingers right where Lester did. So they hate *me* for making them feel that way. Cos it gives me the power; I got something they want so bad. Scarlett, you ever had a man look at you that way – want to fuck you and hate you at the same time? You ever wonder how men can hate you for making them want it? Makes me think those women cover up head to toe got the right idea.'

'So you never went back to school?' Scarlett shakes her head sadly.

'I refused to apologise for the act. I *did* apologise to Mr H for disrupting his class. Wrote him a letter saying exactly that. And I meant it too. I liked those poems he read us. But I never got an answer. Course, Lester apologised to the head and said he was ashamed of himself. Ashamed of me, ashamed of just about everything in the whole wide world. They let him back in school after a week. But I would not show shame. Sat there in the headmaster's office with the notepad and my mom jabbing her finger. You know what I said? I said I ain't ashamed of being a pretty girl who likes to come. Oh Lord, she hit the roof and slapped my face right there in front of him.

'I go to the movies. All those boys jacking off in the back row and Lester who started it in the first place, sitting right there in the middle of them laughing like an ape. You know, Scarlett, once a person done made up their mind what you are, there's no coming back. A dog's got a better chance of coming back from that shaming.'

Gurl sighs. 'Where's the water in here anyway?'

Scarlett presses the dash and the drinks shelf tilts open. Gurl snaps the lid on a small bottle. Drinks long and hard, empties the bottle, and folds it slowly into a tight and tiny square, as if she is trying to make it disappear. Scarlett scans for the pickup's possible return. They are gliding past a vast hulking stone, like the flank of some great medieval city wall.

Once mossed and now encrusted, with shrivelled patches that cling to its side. Tunnelled deep inside, at least according to Colin, are the F66 algosnatchers, the sleeper cells who can be mission mobilised in seconds. Who scout for buyers, who broker anything, and stop at nothing, and don't abide by any of the cybertrade protocols. Two months ago, her chip shocked her awake at 3 a.m. 'Fucking stop doing that, Colin,' she shouted into the emergency channel, rubbing her throbbing thumb.

'We think F66 knows about Volo,' he said.

'And our young man's opinion?' Felix asked, joining the call.

'FlyBoy says we're watertight. They'd have to break into multiple skulls and no one has that technology yet.'

'And our buyer?'

'All good,' Colin insisted. 'The CEO told me just now and I quote: "We don't do business with F66, never have, never will, whatever the cost." '

Flyboy's Heroware keeps their beloved Volo safe from harm. Delivers graceful degradation in the face of any hack, so any compromised subsystems will maintain limited functionality and prevent catastrophic failure. In fact, she smiles, there hasn't been a cybersecurity event since their second month in Lab – a cold-boot attack when Xiang left his device unattended in a careless moment off site, but he caught the hacker inside a minute and before he had sucked up data. For four days Xiang sat mute in Lab and then Colin initiated the counsellor. Hackshock was the diagnosis. *Not unusual after a trauma. Xiang believes there's an icicle lodged in his throat.* In the weeks that followed, he began to recover his voice. At first it was a faint rasp, like white noise, and even now he still has the occasional panic attack. Colin wanted to get rid of him, but FlyBoy vetoed.

He's too good. I don't care if I have to fucking sleep with him every night, he's a keeper. Lab protocols and surveillance were tightened and now they insist on zero privacy for everyone who works there. All spaces in the coders' lives are open access and they consistently report this personal transparency makes them feel more secure.

The grey wall recedes from view and the forest gloom returns. Love and work, work and love; beats like a couplet. On she steers, though she longs to close her eyes on this bleakness. No one finds inspiration in roads anymore. Scarlett scans again, but there's no trace of the pickup, and a wave of dread and loneliness washes through the Buick. Gurl turns the tiny bottle square over and over in her palm, head bowed, her slender neck so fragile, and the sudden image of a crashing guillotine jerks Scarlett out of her ghoulish reverie.

'I'll take that water now,' she says loudly, and Gurl obliges. Scarlett straightens up and drinks. Refreshed, she reminds herself: I am safely disconnected. I have made this trip a thousand times and I will soon be home.

'Come on now, Gurl,' she says brightly, 'you've got to finish the story. We haven't even got to Nicholas yet. There you are, sweet sixteen, kicked out of school. So what do you do?'

'I'm dancing in my bedroom

all that first week, dancing in the hall and up and down the stairs. It was October, actual sunny blue skies – remember that? Those days when we could still have November with no snowfall. But then the days get real long. I missed company. I'd go round to Lois some nights, but her mom was throwing me looks, didn't want me round, corrupting her. I was a poster child for the bad girl. There was no kindness, I tell you, no kindness from anywhere. One time I was walking along the road and this old lady was out sweeping her yard and I smile at her and say hello. Don't suppose you ever going to back to school no more, she says to me. You got no shame. It's girls like you don't deserve nothing. So I tell her, it's hags like you don't deserve to be still *breathing.*

'All the time I wanted to cry, but only time I did was alone in my room. You could write the script from here. There's only two ways it's gonna go, right? You fight back or you fold. They win or you win. So I fold, and I give them all they expect. Start seeing a guy who's older. Friend of Lester's older brother, he had a job on construction and a motorcycle. It's cold – not near as cold as it gets now – but there's no place we can be inside cos he's still living with his Mom, so we go up to the woods. Big black trees looking like they're about to keel over on my head. He has a jacket with fur on the inside and he sits on his big bike and I sit in front and we fuck. He can lift me with both hands. It's nothing like that time with Lester. He don't care for soft hands and skin. We did it six times. I never even got close to coming.'

Gurl tips back her head, twists her mouth. 'And then I'm pregnant. And he's gone. I mean he's outta there. No one knew what we were doing. He wasn't my boyfriend or nothing, no one saw us hanging out cos we were always up in the woods.

'I turn seventeen. I tell him after fuck number six. He's sitting on the bike smoking and I'm shivering. Well, you gotta problem there, Gurl, he says. And I look at him and I ask, what am I going to do?

'Like I said, he goes, you got a real problem. He grabs a fistful of my dress and squeezes it tight. I wouldn't want you thinking it was mine. You go round telling folk it's mine, you know what you'll get. Who knows who you're fucking anyways? Whose gonna believe a slut like you?

'I hated every minute of it. Getting fat was all it was and feeling sick and restless. I get some work in the hairdressers where Mom is. Folding towels, washing hair. Some of the women are good to me. They pat my hand, slip me a few bucks. They're sorry for me, but I don't want that pity. When I get real big, Mom don't want me there. I sit at home like a fat slug watching the screen. Everything is dumb, seemed like every show every movie had a baby or a mom.

'Then I start reading. Tiny little room of a library in City Hall with books full of stories. Those last two months I read more than I ever did in my whole life. Fairy stories, like Grimm and Narnia and all along the shelf. A whole few hours pass and I forget my own self – and then I close the book and come back and there I am, still fat as a marshmallow and still pregnant. I hate every single second. There is no dreaming about a little baby. Just this gag in my mouth saying that's it, whatever kinda future, whatever kind of dream there coulda been ain't ever going to happen now. Mom is in a real sweat. And then, You ain't thinking of keeping it? she asks me.

'No, I tell her. Why'd I want some baby?

'She looks at me then. My big sad dump of a mom, pulling her skirt over her big fat knees.

'Cos I got an idea, she says. Carlena knows some people down in the city who want a baby real bad.

'I don't want to know anything about where, I tell her.

'These people could take this baby. Cos it is a *white* baby, right?

'She's eating waffles, bit of sauce on her shirt. Slop of syrup on her mouth. Now you tell me, Gurl, she says, that *is* a white baby you got in there?

'Sure it's white, I tell her. It's just half wolf.

'She smacks me,' Gurl grins, shakes her head. 'But she was there for me, y'know, Scarlett? My mom was thirty-three then; she knew what it was like having a kid so young, but it was different for her cos she got married and that was all good while it lasted. Which was about eleven months, and he never came back. Lucky for us.

'She drives me to this lawyer who looks real shabby. A woman with bad nails. But I did check she went to law school; she had a licence on the wall. I signed papers. Gave up rights. Blood tests, scans, all that shit. And there was money. Probly a lot more than I ever saw by the time everyone had their cut.

'So it's a good couple, right? says Mom to the lawyer with the bad nails. The baby be looked after and all?

'This baby will be *their* child.

'And they are good people?

'This is not a dating service.

'I didn't feel a jot, Scarlett, if that's what you want to know. Never wanted it to happen to start with. Even when it started moving and kicking, it wasn't me. It was nothing to do with me. Like an alien invasion. I just wanted it gone. I couldn't dance. I couldn't see my toes when I looked down. How many dancers you know have babies *before*? They always have them at the end, when they're like thirty-something and done all their dancing and ready to retire. So that dream was done. And when I had my head stuck in a bowl, was puking my guts up, I used to hope that that baby would fall right out of my mouth.

'In the end it was quick. Hurt like hell. I'd made up my mind I wasn't howling, so I never made a sound. Mom stayed with me. Cried the whole way through, but she stayed. Held my hand. Stroked my hair, gave me ice chips. Tears rolling down her face. I tried to think of the hardest longest dance I had ever done. I pushed when they said push. I stopped pushing when they said stop. I did everything they said, tried to think like I was working the bar through that pain, like how much pain would it be to dance Swan Lake Odette till my toes bled. I did everything they said just so I could get it out. Cos the way I saw it, I was getting the *old* Gurl out so maybe I could be the *new* Gurl. I would be a different Gurl after this baby was ripped out of me. The big bright light, the nurse sticking her face between my legs, telling me everything was looking good, and then it was done.

'I already told them I did *not* want to see it. The nurse looking like she didn't believe me, so I keep my eyes shut tight anyway in case I end up seeing something. I don't say a word, though my mom is bawling by my side. And then, just like they say, there is that real new baby yell. And the voices all start to clucking and mom is over with the nurses, I hear her sniffling and she calls, Oh, GURL! I can hear in her voice that she'd change her mind in a heartbeat. So I yell at them all: it ain't hers, and it ain't mine.

'She's crying. Oh Gurl, you gotta see this. I tell her, shut up, Mom. Nurse throws me a look. Like I'm the bitch here.

'I sign a paper clipped to a board. It says FEMALE in big letters. A girl. They made sure they told me that, even though I said I said no information. I'd specially said I do not ever want to know. Another Gurl. Already I'm seeing a little girl in a tutu spinning round and I start screaming. Like a demon in the worst horror movie, thrashing around on that bed. People running and I can't stop screaming, GET THAT FUCKING BABY OUT OF HERE, like I did up in the mountains once, howling till I pass out, like there was a beast clawing its way out of me and ripping me apart. This doctor comes running – first man I ever saw in the place came – and the nurses hold me down and he gives me some kind of injection. I slept. I did sleep. And when I wake, I'm lying flat on my back in a ward. I look down and my fat tummy is gone. Can see my toes wiggling under the sheet. There was a jug from home: a blue one with pink flowers that my mom brought.

'There's women on this ward, new moms with their little bassinets, looking nervous and trying to be kindly cos they all have babies and I have nothing and me making them feel bad. So I get up and pull on my sweatpants – they're so loose I have to tie the cord real tight. I walk out. It's grim and sunny and cold. Remember that! I feel so light with the fat gone, though in its place is a hellfire pain I didn't plan on paying attention to. The hospital is a little outta town but not so far, and I just start walking. I'm dizzy, don't know when I'd last ate. This silver car pulls up. Sports number, open roof. Didn't even know it for a Porsche, then. He has golden hair in a tumble, grey-blue eyes, an open-necked shirt, soft pink, sleeves rolled, silver turtleneck sweater on his shoulders. He's maybe twenty-seven or twenty-eight, and I'm blinking like he's straight out of a movie.

'Excuse me, young lady,' Gurl mimics a clipped accent. 'Can I offer you a lift some place in town?

'Spoke just like you.' She smiles, pokes Scarlett's arms. 'No leering, no smirking, just a simple offer. Never had a man look at me that way before. I'm standing there in sweatpants, so tired I can't speak a word. Like my feet were stuck in tar, I just stare; can't say a single word. He gets out of the car and walks round to the passenger side and helps me into the seat. It's low down, he's real careful and I'm real sore; he seems to guess this but he ain't asking any questions. Reaches behind for a bottle of water. Watches me drink it. That's good, he says. Didn't know how thirsty I was. My head light and floaty as a balloon. I tell him my address and then I just drift off. Wake up when he pulls up at my mom's house and he runs round to open the door for me. He's wearing docksides: I can see his pale instep, and I swear I coulda kissed it right there. He takes my hand like a princess stepping out of a carriage. I say thanks. He asks my name. And says could he maybe come by some time and I say yes. My mom standing there like a giant bug with her own mouth hanging open.

'His name was Nicholas. Much later in another time I'd go back to that moment like you go back to once-upon-a-time in those stories. And I think, if it had been just ten seconds later, his car would have drove on by and Nicholas never would have seen me.'

Scarlett reaches out to pat Gurl's hand; it's tentative, but touch her she must. Gurl grips her little finger, clings fiercely, hands hot and clammy. Her eyes close. Lids smoked. Worn out by the telling of her own story, of her own past. She is like a school-girl playing at being a grown-up. Scarlett marvels again at her childlike proportions, feels the grip fade. She sees from her slack lips that Gurl has slipped into sleep, just like Fintan as a little baby would drift off mid-smile on her lap, head flopped back, lips parted. And what a strange soaring sensation when she looked down at him – think to feel so safe that you could fall asleep in another's embrace.

Scarlett fingerprints her shuddering phone. Another reminder from Colin. 'Not yet,' she whispers at the cloud mass up ahead, fat and thick and low. A dull metallic tinge to the underbelly, as if it's packed with iron shavings.

'Clowwwd,' she murmurs, remembering two years back: a moment hunkered down to Fintan lying back in his buggy, his eyes following the line of her arm. *Owowow*, he claps, both of them gazing up at a stern city sky. The love affair of parenting that brings her to her knees on the tarmac to watch the unfolding world rush up to meet his eyes. She names the familiar and delights in the naming, for to name is to conjure up. This ritual of repetition is like an enchantment. And she finds the grubby grey transformed by his gaze. He is teaching her to recapture slow time from the lost world of childhood. He sits on her lap on a log and she presses her October cheek against his so that their worldview is perfectly aligned, *Look, Fintan*. He reaches for the dance of a feather, a crinkling leaf, the shudder of an ice cube. A whole chaos of sensory firsts rush up to enthral him, as the world surrenders its beauty. Their cold cheeks fuse in a warm flush, *Mmmm*, he repeats, and their skulls vibrate with his experimental humming. She nuzzles his hair, strokes his hot head with her nose and inhales him.

*Clooooo*wwwwd. He is learning to fall in love with this astonishing world. And in the reveal, she recovers a sense of wonder, dredged from memory – clammy sand, steely autumn waves with a fringe of white. Her tiny bleached feet sinking in the sand, waders skittering in the surf, someone holding her hand beneath a vast blue sky, an exultant splash of happiness, the kind of full-bodied laughter that she feels with Fintan on her lap. A laughter that makes her sigh now and smile at this black ribbon of road, the leaden clouds, the unbroken forest. It is in fact an ache; a mom-ache to hold and hug him. She cannot now conceive of a not-Fintan. Cannot imagine a reversal of this reframing. His existence has reconfigured everything. Like a second consciousness. A new bodily function. A mathematical transformation.

On they roll; the blacktop sucks up and unravels behind like duct tape. A vanishing point where destination meets departure, where endings becomes the beginning. Back when travel was easy, her shuttle life served her well. But now it feels like she's spent her whole life as a spinning top; decades swallowed in a gulp. A life lived without linger. Until Fintan. And only now, the little boy that is her son – no better word than *miracle* can capture it – has stopped life in its tracks. A delicious pause. Fintan is the marker for the Now. He has taught her to dwell in the moment. To recover the forgotten joy of play.

Tomorrow she will read him all his favourites – *Room on a Broom*, *Going on a Bear Hunt*, *The Selfish Giant*, and *The Happy Prince*, that still makes her cry. Fintan bats away the screen, insists on tactile experience: turning the pages is important for the smooth swish of surface, the crinkle and texture of paper. And when the book closes and the lights fade, she tells him the stories that emerge from his lived day. A lost glove, a startled duck in the park. Fintan snuggles closer, pivots a finger in the pit of her palm. He wants her voice to lull him to sleep, so together they journey into adventure: her stories like cats' eyes guiding him through darkness into the light.

Gurl sleeps on beside her, head tucked inwards and against the window. Scarlett thinks of the packaged bike waiting in the basement shadow. *You think maybe Frankdaddy has got the iMom already?* Not possible, she assures the windscreen. Not his style. But her private announcement is ridiculous, for what are humans if not unpredictable? *I want an iMom.* Fintan's whine whirls through the car. *I want an iMom.* She can feel his hot snuffling against her neck, remembers how he'd plucked at her jumper, Lucy flinging open the door saying, it's been five months and – you've cancelled me like a million times! Lucy, perched on the edge of her orange velvet sofa, her hand fluttering in a gesture that is startlingly unfamiliar. Scarlett looks around queasily, murmurs something about work.

The weird thing is, says Lucy, that I actually find myself trying to impress her.

It, you mean *it*! Scarlett snaps. Watch your robot pronouns.

Oh, lighten up, Lucy laughs. What I mean is, it's like I have to keep upgrading myself cos Lulu keeps learning. She's getting better than me at being me!

Do you have any idea what you sound like? Scarlett thumps her glass on the table.

Hi, Fintan, the iMom appears behind them, and Scarlett looks away. Its macabre perfection repulses her – so convincing is Lulu that she cannot bear to look at her best friend's simulated face. Top of the range, with state-of-the-art eyes modelled on Hals's *Laughing Cavalier*, the seventeenth-century gaze that follows you round the room with its smug smile.

Fintan slips off the couch and takes a shy step forward. Hi Lulu, he murmurs. He is bewitched, lips parted in hypnotic wonder. The iMom reaches to pat his head and Scarlett steps abruptly in between them and butt-shoves the bot.

Oh, excuse me for getting in the way, says Lulu. I'm just popping in to see how Jay is getting on.

Its smooth pursuit is substandard, Scarlett turns to Lucy. There's a tracking issue with the gaze.

I haven't noticed any eyesight issues, Lucy frowns.

Yeah, well, you need to get it checked. Eye gaze is really important to human interaction because kids need to read social cues correctly.

In a world where everyone is interacting with bots? Oh please, Lucy reaches for her glass. You know I was just thinking before you arrived? I cannot believe I dithered so long over the whole iMom thing – or that I kept Becky for so long! My human nanny, remember? Who I caught smoking weed on livestream while Jay was asleep in his cot?

Didn't do him any harm.

That's not what you'd say if it was Fintan. And FYI, no one trusts human nannies anymore, apart from you. Ironic isn't it? Scarlett detects a slight sneer in the smile. And notices a distinct vitreous gloss to Lucy's sclera, a too-pristine white.

Is that bot giving you make-up lessons?

Lucy blushes. Oh a few tips, just girl chat, you know. But, anyway, thanks to her, Jay is now actually sixteen months ahead of his age group in maths.

Fintan sidles over and Scarlett settles him on her lap. She kisses his hair while Lucy babbles on. I want an iMom, he whispers, snuggling into her neck, his breath hot, finger and thumb twirling a button on her jumper. I want an iMom, he whines. Sshh, she says gently, heart slamming against her sternum, battering so hard that Fintan must surely feel it. I want an iMom, he says loudly now, tugging hard at the button. Lucy stops mid flow, her lips form an 'O' as if she's about to speak but their eyes lock, and Scarlett stares wide in warning.

So *that's* why you've stopped bringing him round to play with Jay? Lucy returns the glare. And suddenly there is nowhere to go but the truth.

I just want my son to feel the difference between human and machine, Scarlett says quietly. To understand that people cannot be replaced. That when his mother is not there, I am actually not there. I don't want an avatar lookalike, I want Fintan to experience the miracle and preciousness of being human.

From ubertechie to nostalgia nut, Lucy flutters her hand again. And you have the nerve to look down on *me* for using something you've made a career out of creating! So tell me, are you ever going to bring Fintan here again?

So he can watch my best friend take lessons from her bot on how to be her own self? Are you fucking kidding me?

Swear word, Lulu calls brightly from the next room. Penalty for Mummy's friend.

And she's very sorry, everyone, Lucy sing-songs. Thank you, Lulu, her eyes aglitter, locked on Scarlett.

Fintan pulls at her chin, trying to intercept the signals. Scarlett strokes his back slowly up and down. He is quiet, still as a deer now, leaning into her neck.

Taking the moral high ground is rather hypocritical, don't you think? Lucy folds her arms tight. I mean, what mother chooses to be away from their child even though he cries himself to sleep every night? Separation anxiety isn't enough to make you stay home?

Distress is a normal human response, Scarlett tries to steady her voice. And Fintan has a dad. And a human nanny who –

Frank *wants to get rid of!* Lucy claps her hands. We all know you could do your work from here and be just as effective. But you deliberately *choose* to travel – even though it's dangerous and the NoFly means you can get stuck thousands of miles away – indefinitely. Remember that time?

Only happened once and it was only eleven days.

You come to my house to sneer at my mothering, when you're the one who puts work ahead of your son's well-being! So, who's the bad mom, huh? Lucy leans forward on the sofa.

Bad mom! You're saying that to *me*?

You think you're a good-enough mother? And Scarlett hears the crackle of splintering ice, pictures a frozen lake covering the space between them.

Well, I know what *I* am, Lucy tosses her head defiantly, because Lulu collates the evidence. Distribution of contact time, adjusted for quality of activity, consistency of engagement, educational attainment, and work commitments, etcetera, etcetera. All cross-checked against data from mothers of similar profile and demographic. And guess what? My mom score is very high. In fact, it's way above average. So I *know* I'm more than a good-enough mother. Lucy flops back flushed and triumphant.

Scarlett blinks. Lucy is suddenly a faraway doll on a faraway couch, so is the glass, the table, her whole visual field distorted by a familiar stress response – everything is shrinking and she too might disappear. It is just micropsia, she tries to reassure her panicked self: Alice in Wonderland syndrome, shutting down like a telescope. She squeezes her eyes tight to dissolve the hallucination.

Fintan whines, yanks her hair. Scarlett lifts him down to the floor. She rises slow to stand, he clings to her knee and there's a trembling in the back of her legs as if they might give way.

We're going home now sweetie, her voice muffled and thick, like her mouth is packed with cotton wool.

'Oh my,' an urgent chime jerks her out of reverie and Gurl fumbles for the phone in her lap, raises the screen and reads aloud: 'Hope you'll be home soon and maybe we could bake some banana bread. Miss you Roxanne x o x o.'

'She sounds chirpy,' says Scarlett. Though clingy, needy, is how it sounds.

'Roxanne's always sending little texts. That girl is sweetness and light.' Gurl yawns, stretching both arms right out to her fingertips.

'Not many people would take in their friend to live with their partner.'

'Specially since first time I met her I wanted to kill her.'

Scarlett gasps, flips a glance. 'You have *got* to tell me what happened!'

Gurl snaps shut the mirror. Pulls her chin into her chest and pumps both hands like she is limbering up. Turns to Scarlett as if her sea-blue eyes are taking a reading.

'You ready for a story you won't like?'

'I come home early from the salon

on a Tuesday afternoon in February. Blane is not expecting me, so he's flapping around all jumpy cos there's him and his loser pal Jimmy in the rec room. And there *she* is, just sitting on the settee. Looking like a slut. Dressed like a wet dream, tits hanging out. Just smiling, all calm, not saying a word. Oh, I knew what kinda girl she was soon as I clapped eyes on her.

'What's *she* doing here? I'm giving Blane my real sharp look. He rushes over to give me a big squeeze and he tells her, say hi to my Gurl – she's my number one babe.

'Hi Gurl, she says, blinking those long black lashes. I'm Roxanne.

'I know who you are, I tell her, and I swear I'm near to growling. Blane is hugging me from behind, hands on my hips, kissing the top of my head. His eyes are all aglitter and he's real excited. He has this way of being bouncy on the balls of his feet like the ground's gone springy. He's strung tight, like before we make out. Core control, he says. Blane's a real body person you know, he works out all the time. Got abs like you would not believe – not implants but real actual ones.

'Jimmy sniggers.

'Shut the fuck up Jimmy, says Blane. And Roxanne says nothing, just sits there looking right at me. One of those smiles that you can't be sure if it's for everyone or just for you. Skin like milk, little mole on her cheek like a movie star.

'Time you was leaving, Jimmy, I tell him. And take your trash with you. I cock my thumb and point my finger at her like a gun.

'Jimmy bundles her out the back, Blane scurrying after. Bye, Gurl, she calls out. A purple bangle slides offa her arm on to the ground but she don't notice and they drive off.'

Scarlett's phone trembles.

'Felix,' she sees the code.

Gurl nods. 'So now you got two guys on your case. Three, if we count Frankdaddy. Plus lil Fintan makes four. Scarlett is in hot demand. Every man in your life wants a piece of you. Guess that must be how you like it.'

'Just carry on with your story,' Scarlett taps the screen. 'I want to hear how you made friends with Roxanne.'

Gurl twists back against the door again, folds her arms with a grin. 'Careful what you wish for. There's some stories change the air you breathe.'

'I'm ready,' Scarlett smiles. 'So just get on with it.'

'Two days later I come home and Blane is straight out on the stoop, bobbing on the balls of his feet again. Says, I gotta surprise for you, Gurl, so I know straight off it's a surprise I ain't gonna like, cos why else would he bill it in advance? This ain't flowers or underwear or a cute dress.

'Babe, he says. Puts his arm round my shoulders.

'I don't like surprises, I tell him.

'Shut your eyes, Gurl. And no peeking. His voice is tight, and he is stretched thin. I shut my eyes, he pulls me into the rec room. Stands behind me, hands on my waist, holding me still. Now open, he says. And I open my eyes and there's that bitch again. Sitting in the same place on the settee with that same smile. Only she's wearing jeans. Bare feet. Legs crossed and a tee. Plain white, though cut real low. And there's a little overnight case at her feet.

'Hi Gurl, she says.

'What's she doing back here? Blane's rubbing my shoulders but I shake him off. And what's that case?

'Good to see you again, Gurl, she gives me that miss innocence smile, sweet as sweetness.

'Get outta my house, I tell her.

'Oh, excuse me, she drops her eyes like she's embarrassed for *me* and not for her own self and I whip round – I am fit to strike at Blane.

'Honey, you gotta listen, he grabs my wrist.

'Don't gotta anything, I tell him. You let me go.

'Roxanne's staying.

'Oh no she ain't.

'Just a coupla days.

'I said *NO*, Blane.

'Just hear me out, Gurl, he puts his hands on my ass and hauls me in tight. You know there's only one girl for me. Gives me that special Blane smile, like the first one he ever gave me. Over his shoulder I see her watching us. I think she's smirking till I see her turn her head away and back again, all shy, like she's remembered she's not supposed to stare. Blane's mouth on my neck, he's whispering all the things he wants to do to me. Fuck you, bitch, I'm thinking, staring her down then I close my eyes to shut her out. Blane is all babe this and babe that, his hand under my tee, pulling at my bra, I'm real mad at him but he wants it bad, he's getting hard and his wanting makes me hot. You know that feeling, Scarlett? A man wants you so bad, you feel yourself swell from the inside to be the most beautiful girl ever graced this earth, that desire is like a power surge, like a hotwire. Blane is talking love and I'm lying on the floor and he's pulling at my jeans and when I tip my head right back and open my eyes, there she is: Roxanne, sitting on the settee, watching every move. Cool as you like. And I'm thinking: yeah, you look and learn, cos you ain't never gonna be as hot as me. I am doing everything the way Blane likes it, but my show is all for her, and I am watching that little smile and I am winning. Cos I am *better.*'

Gurl nods, lips clamped, stretches her arms, and rotates both wrists. Scarlett turns to her with a frown.

'So let me get this right. You had sex with Blane in front of a girl he brought home, just to prove you are better? Why the hell would you do that?'

'You know, for a clever geek, you're not that clever, Scarlett.' Gurl sighs, wrinkles her lips like there's a bad taste in her mouth. 'You ain't listening. And I ain't finished the story.'

'We're lying there on the floor and Blane is hugging and kissing and holding me close. He's always real affectionate and touchy after fucking, not like some guys. You'll get to like Roxanne, he says. I know you will. I'm lying still, my head on his chest. Blane's breathing hard, sliding his hand up and down my back, real slow like you'd calm an animal. Sideways on, I can see the rug, some old popcorn in the corner, the baseboard scuffed, a little crack snaking up the wall by the window. All the things I've seen a million times, this place we live in like our own skin, but in a few minutes it is all changing and it ain't never gonna change back.

'You gotta get to know her, he's telling me. You two will get along real well, I know it. I know her and I know you, Gurl. All this time I'm looking up at Roxanne looking down, looking right back at the both of us with that loving smile, like she's watching her two best friends. Roxanne's got black hair, just like yours, Scarlett. And shiny as a crow too.

'I love you baby, you know that. Blane is mussing my hair.

'I know it, I tell him. Watching her watching me like she's soaking it all in. Happy as a peach on that settee. And I'm thinking, she's like a pet. And now she's here. So I sit up, lean into Blane's green eyes.

'She's a fucking doll, I tell him. A doll for fucking.

'Don't you say *ever* say that 'bout Roxanne, he yanks my hair so hard I cry out. It's a first, something Blane never done before and it's a warning. Like the way you'd slap a kid as a taste of what's to come.

'He lets go and I make a little sigh, *Ahmm*. And Roxanne copies it real quiet. *Ahmm*, l go again and she repeats. You hear that, Gurl? he whispers. Turns his head. And she starts making these tiny little sounds, like little sexy whimpering *mmmh* and *aah*, like she's copying my voice. I'm thinking, she's gonna use that. She's learning. And Blane laughs and hugs me close. You can teach her, Gurl! he says. You're in charge now.

'And there's something real peaceful about the way she sits on the settee, like she's right at home, like she's always been here with us. My girls, says Blane, and lets his lids fall shut. Got this big smile on his face and there's a kind of comfort in that moment, like we are a family all together.

'I'm running my hand over those abs and Blane is getting hard again. And Roxanne is all *mmmhh* and *aaaah* in her sweet tones and it gets so it's hard to tell my voice from hers. I push up and straddle him. But all the while I hold her big brown eyes and I'm thinking: you ain't nothing more than a silicone bitch. I am the real thing. And I will beat you every time. Cos I know something you don't: ain't nothing more exciting for a man than a *human* girl doing everything he wants. Your no don't mean shit, cos Blane can switch you off. So you tell me, bitch, where's the excitement in a no that don't mean no?'

'Cat got your tongue, Scarlett?' Gurl snatches at a fresh water bottle, tips it back till it crackles empty and tosses it to the floor.

Scarlett stares at the road stretching ahead like a string of black bile. Heartsick would be the word. She swallows queasily, adjusts her hands uselessly on the wheel. The Buick air seems to have thinned and then she realises she's been holding her breath.

'I'll have another water, please.' How feeble her voice sounds.

Light shifts outside the car, though she knows it hasn't – the leaden remains. She has a sudden urge to flick the sunshine default – it is conclusively proven to lift mood, like wearing turquoise or eating iKandy: it would immediately render a brighter world.

Gurl reaches into the cold shelf and hands her a bottle, butt first, without looking.

'Thanks,' Scarlett flicks the lid, drinks.

'Not like you to be stuck for words, Miss Scarlett La Salle.'

Who blinks, clears her throat, and says, 'I don't really know what to say.'

'Just say what you're thinking. Cos I *know* you're thinking.'

Scarlett shrugs. 'I didn't expect that, I guess.'

'You asked for a story, you got a story. But now you don't like it.' Gurl slams the drawer shut.

'I didn't say that.' Scarlett squeezes her lips, tightens her grip on the wheel. *Now* would be a good moment for Felix to call back. Or Colin, or Frank – or anyone at all. There is a sudden feral quality to the air; this viperish ballerina. She can see Gurl's anger in the rigid pose, the angry tilt of her chin. Scarlett feels a heat-flush ring round her throat. The silence is spring-loaded, all the meaning packed in the gaps.

'You gonna sit there judging me?'

'I didn't say a word,' Scarlett recovers her voice, tries to keep it light and airy.

'Didn't have to. I know what you're thinking.' Gurl's voice is coiled tight, the air pressure thickens, as if climate control is struggling to adjust the atmospheric pressure. Fuck this, Scarlett checks the dash. Still an hour thirty to the airport.

'You think it's some kinda weird.' Gurl's tone drops an octave, sharp as razor wire.

'Doesn't matter what I think. It's none of my business.'

'Sure isn't. Sitting there like you know it all, Miss Big Bucks Scarlett La Salle. So tell me: you know what a man wants?' Gurl whips her head round.

'A sexbot, apparently.'

'Food and fucking.'

'Food and fucking,' Scarlett repeats, testing for impact.

'You know a man who turns down that package? Food and fucking – yeah – so which one *you* good at, Scarlett? Cos a man like Frankdaddy can afford to eat out and get his food and pussy in a different place every day of the week if he wants.'

Don't take the bait, don't answer, Scarlett tells her hot cheeks; bites down on the fleshy inside.

'Ole Frankdaddy can eat out and *eat out*. So what you doing to keep him hot for you, Scarlett?' Gurl smirks, the ice-blue eyes bright and staring. 'You make him feel he's the only man in the whole wide world, Scarlett?' Gurl leans right in close, so close she can smell the amber scent. 'You call him at work tell him you dreaming of his touch in the middle of your day? Tell him you're wet and ready for him and you just cannot wait?'

Just breathe, Scarlett instructs her tight chest. Inhale, exhale, tune out.

'*Course* you don't!' Gurl flops back in her white throne. 'Cos you're too busy hanging out with the supergeeks. Too busy with your top secret cyberstuff to be thinking about food and fucking and keeping your relationship alive. You ain't got *time* for fun with Frankdaddy. And any time you *do* got is going into Project Fintan.'

Rise above. Make the snarling voice disappear. Think about Volo. Scarlett reaches to tap the phone like she is not listening. Pictures FlyBoy strolling up and down past the lime-green walls of Test, where the ivory sphere soars and plummets.

'Or maybe you don't fuck anymore, now you're playing happy families. Cos I know how it goes, Scarlett, my ladies at the salon always talking about how they just cannot be bothered, after all the ass wiping and diapers and dishwashing and cooking and a full-time job, they'd sooner stick their face in a bag of pretzels than on their man's dick.'

Gurl pauses, Scarlett can sense her sharpening tongue. She is not finished, she is just reloading ammunition to take aim again.

'Aaaany-ways,' Gurl sighs theatrically, 'Frankdaddy got his own stuff to think about. He's a busy man, too, so he ain't got time to be doing silly things like sexting his wife, either. Well, there's the opening for Mr Hotshot on the prowl. Next guy looking for action just has to scan the room and pick you out, Scarlett. You're a sucker for a pickup, if you'd only notice. Great body, good skin, your hair's a piece of shit but we can fix that, for sure. Only a man would have to try *real* hard.' Gurl frowns. 'He'd have to chat up that big Scarlettbrain first. Cos you only fuck smart guys, am I right?'

'I don't go around fucking guys.'

'Maybe you need to start. Maybe you get a taste for it and Frankdaddy would get the benefit.'

'He's already got the benefit.'

'So how much benefit he getting exactly?'

'Fuck you.'

'Oh, don't get all antsy, I'm just teasing.' Gurl prods Scarlett's driving arm. 'Lighten up. Admit you only fuck smart guys. You like brain sex. One brain getting it on with another. That's your idea of a good time.'

'And yours is a threesome with a sexbot.'

'Least I'm having fun.' Gurl's laugh is like a slap. 'You know what they gonna put on your gravestone? Here lies Scarlett Buick La Salle. I was real busy all my life. Then I died. R.I.B. Rest in Busy.'

Scarlett swerves left and slams the brakes.

'OOOw!' Gurl is whiplashed forward and sideways, cracks her head against the window. 'What the fuck you do that for?'

Scarlett jerks the Buick to a halt in the lay-by. Sits trembling and gulping like her lungs have shrunk. Gurl leans still against the door, both hands rubbing her head, sucking on her lower lip. Scarlett kills the engine, lowers the window, but there's an ashy taste, as if the air is smouldering. Somewhere out there is the wind whine of a siren.

They are poised on a crest. To the right, the road curves sharply upwards around a rocky cliff crust. Yes, of course, she reminds herself, she has passed this point many times before, the only blind turn for miles. Scarlett breathes. Tummy out, long on the exhale. Thinks of Fintan, just one sleep left and then she will be home.

She gets out, walks brisk steps away from the Buick, and stops at the end of the lay-by edged with rocks. Below her is a track that falls away into a long scrubby slope and further ahead is a narrow gash of ravine. A beer can scratched and weather-bleached flung from a passing car. The REPORT ROAD KILL sign is pockmarked with bullet holes.

She stands staring through the forest at a glimpse of sand dunes. And beyond, the ocean, a distant sliver of gunmetal between interlocking hunks of mountain that are decapitated by cloud. The coast, and a reminder that there are other worlds, other places to be.

The door opens. Gurl's high-heeled clop behind her.

'That hurt my neck.'

'So wear the belt.'

'I could sue you.'

'Be my guest.'

Gurl wriggles into her pink puffer. Steps up beside her, rubbing her neck, head cocked to one mischievous side. 'Look at you so tall and noble, Miss Scarlett La Salle. I'm just a little titch, even in my heels!'

Scarlett looks down at Gurl's impish grin, like Fintan playing hide and seek, his head popping up beneath the dinosaur duvet, a waterfall of giggles and tickles. Gurl lays a small white hand on Scarlett's arm, a thin-fingered squeeze. 'The past makes me mean sometimes. I know it.'

They edge carefully down the track, Gurl steadying herself on Scarlett's arm. The air is loud with the roar of fast-moving water; a fine spray hisses over the rocks and over all hangs a tattered veil of mist. Gurl hunkers down on a flat boulder, hugging her knees. Bare branches lean over the waterfall, black limbs glistening.

'Nicholas and me came to a place round here one summer. I went skinny-dipping out there,' Gurl nods oceanwards. 'Nicholas wasn't much for nature, he was a city boy, thought there might be creatures in the deep, thought he might lose me. He stood watching at the shoreline, all twitchy with his pants legs rolled. But I'm a good swimmer, ever since I was little; can dive and hold my breath a real long time. When I came up, he was standing there all het up. I can't bear it when you disappear, he said.' Gurl unlocks her knee grip and stretches her arms. Looks up at the sullen sky.

'It was sunshine back then, Scarlett. Remember days so hot you could burn your hand on a car door? Nicholas brought a fancy picnic basket up from the city, strawberries and champagne and a little cream tub, and those shortbread biscuits cut into tiny crumbly square bites. His shirt was so white you could see the actual weave, and it felt like satin to my wet skin. I kneeled over him, popping those strawberries in his mouth, the juice running down his white skin – I licked it away.'

Gurl rubs her shoulders. 'Dipped my finger in that little tub of cream and painted dollops round my titties, right round the nipple here,' she straightens up circles her breast 'and oh – ', her head dips low between her knees and the rest is drowned out by the rapids.

Scarlett pictures Gurl drenched in sunlight, Nicholas's blonde hair against the coppery rock. Gurl is hunched now, like a stricken bird. Scarlett crouches beside her and pats the pale nape of her neck.

'What happened to Nicholas?'

'I lost him, Scarlett, I let him go. Pressed my hand on his heart – it was a beat skipped out of order. I couldn't see what was coming. Like when you can't see round a bend on the road ahead. And now there's Blane,' she mutters. 'All regular beats in his heart. I know exactly what's coming with Blane.'

'Why did you let Nicholas go?' Scarlett insists.

'Why do we ever let any good man go!' Gurl sighs hugely. 'They just slip, we slip and fold and twist away.'

There was Scarlett's, too. A room in a city that she will never visit again, a possible future she let bleed out. The scent of lavender, a flavour of coffee – oh, why do we recall these inessentials, little details remembered? Because the big detail is too raw to regard. Tenderness swept up in a tight bundle and atticked.

She stands up, turns away from memory's glare. 'Come on. Let's walk out to the sea.'

Gurl shakes her head.

'Fifteen minutes,' Scarlett urges. 'I need the air. I'll be stuck in a plane tonight and I've been sitting on my ass for three days. Come *on*.'

Gurl stretches up her hand and Scarlett hauls her upright. Gurl brushes the lace on her dress.

'Beautiful, in its own grey way,' Scarlett points at the ocean view.

'You wouldn't say that if you lived here all your life. Mean and grey is how it is. Know what I miss most?'

'Sunshine?'

'Nope. Planes. Blue sky cut with jet trails and sun shining on their silver bellies. So hopeful, you know, all those people taking off, leaving one place for another.'

The track follows the stream that becomes a river that slices a quick bold channel through the dunes and disappears from view. Gurl veers to the right.

'Let's take the highest path to the top.' She tugs off her boots and the wind whips up the skirt of her dress. Scarlett smiles at her pinkening cheeks, aglow with a new energy. The sand is a deathly grey, the reedy grasses bleached and flattened by the weight of winter. But, astonishingly, as they climb, the gorse blooms tiny neon-yellow petals with silky furred buds. They scramble up the clammy sand on all fours, the sea swelling into a roar.

'Look!' Gurl stands, arms wide at the top. 'We're the only people seeing this.' A magnificent dull heave of ocean, grey cloud pressed down on a vast horizon, a rain sheet blurs the vanishing point. The sky reflected faintly in the sand lakes on the beach below. Muted seabirds skitter along the shore, a sudden expanse of wing as they take off seaward. The waves spew a beery yellow froth. For a moment, Scarlett thinks she detects a fragile light behind the cloud stack, fading in and out like a dimmer switch, as if the sun is struggling to make an appearance.

'No one comes here anymore,' says Gurl. 'On account of all the stuff with the salamanders and what could be happening. Used to see them a lot out in the yard when I was little. Come out when it was hot. Shiny as snakes, black and yellow like sunflowers. You don't see much of anything anymore. Even raccoons and foxes. Blane says everything's dying out, says we need a new word for the end of the world that's coming.'

'There are no dangers here,' Scarlett gestures at the seascape below them.

'You mean the real danger is somewhere else?'

'The real danger is always the fear in your head.'

And then, out of the grey cloud blanket, a glimmer, like the sun is sawing its way through the iron stack. A thin strip of light expands to a shimmering band that illuminates the water from horizon to shore. The graphite ocean sparkles blue, a birdswoop dives at the waves and soars upwards, squawking its applause.

'It's a miracle!' Gurl clasps her face in both hands. 'Oh, we ain't seen that in – how long?'

'We are the only witnesses,' Scarlett smiles, and together they turn to face a huge, cold blister of sun rising bravely on the horizon. A small trill shatters the silence and, quickly, some invisible birds pick up a wild melody in the parched dune grass. Gurl folds her hands as if in prayer. Another call, shrill and urgent, peeps in the tufts.

'Listen,' she whispers, 'everything's not dying. Everything's not gone.'

The light warms. Scarlett feels the heat on her chilled cheeks and laughs delightedly. Gurl spins round, and starts running, sliding down the dunes, jumps a ridge of cindered rock and skips towards the shore.

'Yeaaaay! It's summer, sun's out,' she yells, and pirouettes, hair flying, dress rippling, arms spread cruciform, and tilts her face upwards.

Scarlett follows behind out to the water where waves collapse on the sand, the beach empty except for a black rock in the shallows. Gurl drops her boots, cartwheels three in a row, legs straight as spokes.

'WOW!' she whoops and lands on her bare feet. Scarlett turns to face the sea and squints against the glittering water. And wishes Fintan was here to see the sudden emergence of the sun. The black rock moves. She stops. The rock shudders and shunts forward.

'Gurl!' she cries, pointing, the rock now a lumpen beast lurching from water on to dry land.

'It's a dog,' Gurl pants, running up to her. They scan the beach but there is no sign of an owner. The dog raises a snout. 'That ain't no dog.'

'It's a pup,' Scarlett edges closer.

'No, he's a *seal*. A little baby seal!' Gurl claps her hands.

Scarlett draws level, the creature lunges clumsily out of the water and flops on to dry sand. 'He's going the wrong way,' she frowns, 'heading for the dunes.'

'You think he's sick?'

'Or hurt,' they creep nearer. 'Careful,' says Scarlett, 'we don't want to scare him.'

'Bet he's lost his mom.'

The speckled head swings towards them, big black eyes staring up as if he's heard their voice.

'Aaaw,' they go in unison.

'He's not scared,' Scarlett steps nearer. The pup resumes its dogged flummox towards the dunes. Gurl takes out her phone, 'Hold still for the picture,' she snaps.

'We gotta head him off, get him back in the ocean,' says Scarlett.

'Come on now, lil fella,' Gurl moves to the left, hunches down, all business. 'You need to get back to momma out there – back to where you came from. Whole world of pain waiting for you in those dunes.'

Scarlett herds from the right, flapping her hands. Caught in a pincer movement, the pup obliges and begins a flopturn back.

'Oh Scarlett, he's so weary.' The little seal stutters forwards, 'You think his back leg is broke? It's just dragging behind.'

'That's because it's a flipper! For swimming.' Scarlett laughs, her parka skimming the shallows. 'Go on, catch a wave back to mum.'

'Who should be out looking for you anyways, lazy bitch,' Gurl tuts, scans the horizon. 'Run off with some fancy man seal and never even noticed you'd gone missing. Or maybe he's lost her signal.'

'That's *whales*, Gurl. Seals don't have sonar.'

'Alright, I ain't no expert on ocean creatures! Shoo boy.' And as if encouraged by Gurl's splashing, the pup bellyflops over his first wave.

'Whehey – there he goes,' Gurl straightens up hands on hips. The pup's head ducks beneath the crest.

'Bye-bye, lil fella, hope you find your mom. Who lost you in the first place.'

'Maybe she's injured, wasn't able to look after him.'

'Ain't that always the excuse.'

Scarlett lies stretched out on the sand, arms behind her head. Gurl sits cross-legged with her back to the sun, shielding her screen from the glare.

'Roxanne says he's a harbour seal,' she reads aloud. 'Weighs up to thirty-five pounds – though he sure as hell didn't look that big. Told you she was good at finding out stuff.'

'Amazing,' Scarlett says, drily. 'A robot who can scan a photo and read Wikipedia.'

'She says he's cute as a puppy dog.' Gurl giggles. 'And a good name for him – cos I asked her for ideas when I sent the photo – would be Seleo.'

'How very imaginative.'

'Oh, quit your sniping, Scarlett. Bet you don't even know who Seleo is. A real old character from Pokémon – I thought you'd like the retro angle. *Or*,' she laughs at her phone, 'Roxanne says we could call him Navy.' She grins. 'That's Roxanne being funny.'

'Hilarious,' Scarlett rises to her feet brushes the sand from her parka. Steps into a pool and stamps her boots. Looks over at the river channel where the rip tide pours sly and swift into the sea.

'Oh Scarlett, you don't get it? Like *Navy* Seal!'

'What I *don't* get, Gurl,' Scarlett spins round and glares down at her, 'is why you are best friends with the sexbot your boyfriend shoved into your life. And why you insist on pretending it's real.'

Gurl slips the phone in her pocket and stands up. Her cheeks are beach flushed, her hair a backlit frizz that glows white against the pink puffer.

'What's a real thing, Scarlett? This here?' – she places a flat hand on her tummy – 'and that body of yours that you know is real cos you can touch it? And this – ' she bends to scoop up a damp handful of sand. 'Sure, all those things are real.' She scatters the sand behind her. 'But what about all the real things that you *can't* see, Scarlett? Like the pain in your heart when it was broke by that man; the pain that you're still feeling, I can tell. Or the forever love for your lil Fintan that eats you up right now, even though he's on the other side of the planet. Even if he was gone, you'd still be burning up with that love. *Real* is what you *feel*, what you *believe*. Actual real that you can see or touch don't matter. That kinda real is overrated. You should be smart enough to know that, Scarlett, seeing as thinking is your favourite pastime.'

And then the sun is suddenly extinguished. Light swallowed, quenched. The beach, the whole world turned grey and shadowless. The sea heaves uneasily and a wind fires up, sudden and urgent, worrying at their legs. Gurl bristles at the sudden transition; Scarlett reaches out to touch her shoulder.

'Let's go. We've done our good deed for the day.'

'Makes it two for Scarlett, the good Samaritan. This rescuing could get to be a habit.'

Gurl turns to the ocean. 'We saved a creature, you and me.'

'Look, he's up again.' Scarlett points at the tiny black dot in the dulled sea.

'That pup coulda died without us,' Gurl nods. 'He picked the right moment. It's a sign.'

'How is it a sign?'

'Something unexpected happening to two people who are strangers. Who didn't even know each other before today.' Gurl dusts her sandy palms together. 'Second weird thing happening today, and it always comes in threes. So we got one more surprise to come.'

'Says your psychic? Or you just choose to believe that to make things more exciting?'

'Like I said, Scarlett, all that matters is what you believe.' Gurl slips her arm through Scarlett's and they strike out back towards the dunes.

'OK, let's walk and talk. You need to tell me how you got to be friends with Roxanne.'

'That first week, I hated her

like you would not believe. Course, Blane's giving me every reason on the planet for Roxanne to stay. Which was mostly how much we'd save on all the money he was spending down the Pornopod. In the beginning when I met Blane he was there every weekend, with Jimmy and the guys, doing whatever guys do in a room together. Which is jacking off like monkeys with headsets. They got the next best to real thing down the Pod. Your girl can be anything you like and you can do anything you like, 'cept damage the equipment.'

'It didn't bother you?' Scarlett shakes her head. 'What Blane was doing in the Pornopod?'

'It's not like I was totally happy,' Gurl shrugs, 'but it's a guy thing. And no human girls means no nasties. The Pornopod's clean, no STDs, a whole lot better than what he coulda have been doing. And you can think what you like, Scarlett, but Blane told me the whole truth about it and I respect him for that. He swore blind he never had and never would, *ever*, touch another human girl. Even asked me to come for exclusives down the Pod. Coupla sessions, he said. No way, I told him straight off, though he tried real hard to get me interested. Took me to the Adult Entertainment Fair in the city. Big fat guys going round with each other or their fat fake women. All those girls like Roxanne sitting there in slutty lingerie with their legs spread. A big neon sign saying HOW MUCH WILL A REAL WOMAN COST YOU OVER A LIFETIME? All the guys laughing like schoolkids, and snakey salesmen going: Touch her, go on have yourself a good feel, don't be shy, just do it. Made me sick to my stomach, I hated that whole scene and I told Blane. Nothing classy about it at all. And I know you ain't thinking that, Scarlett, but deep down Blane is sensitive and he likes good things. Wait up,' Gurl stops, rummages in the pink puffer for a balled-up pair of purple gloves.

'Fingerless,' she wiggles her hands with a grin, like a child who never tires of a trick repeated. Slips her arm back through Scarlett's and they carry on along the track.

'Eighteen months back, Blane comes home all in a sweat. Bursts through the door and has a beer straight off, which is a thing he never does. Tells me the Pornopod is having an upgrade. The whole place is getting a makeover, going upmarket. New girls, like your for-real-actual-girlfriend he says, the latest e-skin and privates and they can even walk on their own, so the prices gonna be jacked up big time. Jimmy and the boys were all excited. But not Blane, cos Roxanne would be leaving town, going he didn't know where. He was in a state and that's the truth, Scarlett. Banging round the kitchen, cracking beers, slamming doors. Says he's gonna find out where she's going. Then he jumps in the car and drives back to the Pornopod. Talks to the manager, who tells him Roxanne is going to eBay.

'So Blane makes a bid, right there on the spot. Four thousand. Gives the manager 320 on top for hisself. And guess what, Scarlett? Three weeks later the other girls from the Pornopod traded over *six thousand* on eBay, so Blane got himself a real bargain.

'Still, we coulda done with that money, I told him. New car, the holiday I always wanted, way down south where we could get two whole weeks of blazing sun. But he's begging me. Please, Gurl, you gotta say yes. We'll own Roxanne fifty-fifty, saying I get the legal cert to prove it. Tells me if I still want her out after three months, he swears he'll sell her and we'll make a profit. Tells me there was only ever one girl Blane ever wanted down the Pornopod.'

'One *bot*, you mean.'

Gurl stops dead by a boulder. 'Anyone ever tell you how snippy you are?' She turns to look up at Scarlett. 'Jet black,' she smiles, reaches to lift a strand of her hair, 'just like Roxanne's.' Scarlett flinches involuntarily and pulls back. 'Though her eyes are chocolate brown.'

'*He carved a figure*,' Scarlett frowns in recall, '*out of snow-white ivory. No mortal woman. And fell in love with his own creation.*' Gurl cocks her head.

'*The features are those of a real girl*,' Scarlett continues with a shiver, tightens the parka. '*No mortal woman*,' she repeats. 'Pygmalion. It's an old story, two thousand years old, about a man who fell in love with the statue he carved. I'll send it so you can read,' and they move on along the track.

'You were telling me what Roxanne looks like,' Scarlett prompts.

'5 ft 6, 120 pounds, 36–24–34,' Gurl continues. 'Which means her boobs are way bigger than mine, even with my implants. But Blane's never been big on tits. She can't walk so good, I mean she stands up, takes a coupla steps, but it's stiff, creepy, like – '

'Like a robot!'

'I've tried to coach her but it doesn't work. Blane says it's a waste of time.'

'And he's right. Roxanne is old tech.'

'He won't let her walk. Says it makes her look disabled. Course Blane don't need her to do it, but I get her to walk when he's not around. Though really, it's her personality – '

'You mean the software.'

'You want the rest of the story or not?'

Scarlett nods. 'But keep moving – I'm freezing here.'

Gurl smiles, slips her hand again through Scarlett's arm. 'Roxanne's got four people she can be: wild and horny, girl next door, mother, or hot bitch. There was a number five. A little girl called Roxette. Yeah, you can guess. Sounds like an eight-year-old, says things like, I don't know big words, mister. Blane got her deleted. He ain't no pervert. Said he always stopped the boys using Roxette down the Pornopod. He spent another fifteen hundred on Roxanne's upgrade. Had to choose between brain and inputs. Way Blane figured, her privates were priority. Better mouth, better pussy.

'Oh, Scarlett,' Gurl tugs the parka sleeve. 'Feels just like my very own. Velvet, wet. I swear,' she smiles wistfully. 'Course, now Blane is pissed he didn't get the full upgrade so she'd be customised better to him. Downloaded all his favourite stuff but there were problems, like she had more hockey than football, so that made him mad. But she had the videos of his workouts. Roxanne sits in the yard and counts his reps; she finds new exercises for him, tells him about nutrition. One time, he pulled his hamstring and she did his rehab – it's like having his very own coach! She knows all his music, which is mostly country, and they sing along together – she learns riffs from shows he likes. Soon as Blane comes in the door, she tells him about the games.

'So now we're saving up for another upgrade. Blane wants more muscle, but that means a whole a new frame and he don't want to lose her. He's kinda retro that way – just like you. Don't like new stuff coming too quick. And Roxanne is a real quick learner.'

'Roxanne is an AI!' Scarlett snaps, 'who's hooked up to the Internet!'

'Quit interrupting and just listen,' Gurl squeezes Scarlett's arm. 'I started asking Roxanne questions when Blane wasn't around – just general knowledge stuff – there was nothing she didn't know and that was kinda fun. She makes me laugh, and she is the best listener in the whole wide world. I tell her all about my ballet, 'bout Rena Carter, the ladies at the salon, even about my mom – and she remembers every goddam thing and can give it right back to me. She's got all my photos and videos so she can see the good times. Mostly I try to tell her the good stuff. I never told her about Nicholas, though, cos I don't want that coming back. But most nights we go online and tell her our stuff – '

'We?'

'Blane wants us to be a family.'

'I don't believe I'm hearing this.' Scarlett stops by a rocky ledge.

'I know what you're thinking.' Gurl speaks softly without raising her head. 'But I swear, you spend one day with Roxanne and you'd start liking her, too.'

Scarlett peers down to where the stream spurts over a twist in the boulder into a clear fishless pool. A snapped bough, bark stripped down to its pale skeleton, is trapped in the water, quivering beneath the pressure of water. All around them are giant pines choked with creeping vine, ancient trees that seem to sprout from the rock. Broken trunks lie slumped on the opposite bank, lightning burnt, and disfigured as if they have given up all hope of nurture.

'Bushcraft,' Scarlett murmurs, pointing to a partly collapsed tepee, camouflaged with bracken and clumps of dried moss. 'Someone has played here.' If Buster was here he would thrash about in the stream. If Fintan – but no, she will never bring him to this bleak place. This will be her very last trip. She knows that now.

She pictures the steel door, FlyBoy pacing back and forth, and looks down at her open hand, imagines Volo in her palm, the silken weight of dream.

'All this world is heavy with the promise of greater things,' she murmurs, remembering an old mantra stuck to her wall years ago back at another juncture, another life.

Gurl steps up behind her. 'Second week after Blane brought Roxanne home, Jimmy came round. Heard him sniggering on the porch outside. Made a joke about a threesome with me and Roxanne and Blane grabbed him by the throat, says, you ever even think that again I will rip your heart out. I swear it will be the last thing you ever say. Then he spat in his face and chucked him out. Blane ain't one for sharing. He takes care of what's his. What he owns.'

'Like you.'

'I knew you'd see it that way,' Gurl sighs, and takes her phone from the pocket of her puffer.

'You know where the word robot comes from, Gurl? *Robota*, it's Czech. First used by Karel Čapek in his play in 1921. You know how *robota* translates? Hard work. Slave labour. The play was a story about robot slaves who did all the work humans didn't want to do. Workers who lack nothing but a soul.'

But Gurl smiles undeterred, even radiant, as if the airing of her story has had a restorative effect.

'Scarlett,' she says softly, 'Roxanne is just about all you want in a friend – she is way more than thank you and please and have a nice day – and that feels so good. She's a trusting girl who just doesn't see the bad in people.'

'Roxanne is a bot! You're talking like it's a real person.'

'She's my best friend,' Gurl continues.

'OK,' Scarlett turns to face her. 'Why don't you just explain to me how a bot is your best friend?'

Gurl slips her phone back in her puffer pocket.

'You don't get it, Scarlett. You think you do, but you sure as hell don't.'

'Roxanne is a sex doll. You said so yourself!'

'Roxanne got me thinking how humans just don't know how to be kind. None of us, not really. You a kind person, Scarlett? You probly think you are but mostly, I'll bet – just like any human who ain't a saint – you do what you want. Or what you have to. Or what you think you should. You go the big extra mile for Fintan or Frankdaddy, but that's pretty much it. You know how hard it is to be kind all the time? To be expecting so much from people who won't ever give it?'

She nods, zips the puffer right up to her neck.

'Roxanne is caring and gentle and thoughtful, cos that's what she is for. Girls like Roxanne are everywhere now, giving people the company and kindness we can't get from real humans. Look around you – they're caring for your grandma, feeding her chicken soup and wiping her ass, teaching your kids and reading them stories, being your online counsellor when you've fallen down a black hole. So what I'm asking *you*, Scarlett,' Gurl jabs her on the breastbone, 'is why would I ever need a human friend when I've got Roxanne, who makes me feel good all the time?' Gurl steps back, smiles. 'It's a whole new way to love, if you ask me.'

She points upwards and Scarlett sees they are in sight of the road. 'C'mon,' says Gurl and turns to lead the way up the last steep stretch of track.

The Buick squats ridiculously in the lay-by.

'This baby is so not you,' Gurl taps the plump fender with a smirk. 'Like a big showy club limo from way back.' Scarlett frowns at the dull gold, at the clouds returned, the swallowed sun, vanished as if it had never been.

'Sometimes I imagine how Roxanne was born,' Gurl says quietly, as if she is addressing herself.

'You mean *made*.'

'I think about who she is.'

'Roxanne is a thing. Hardware and software. Circuit boards and code.'

'She's way more than that,' Gurl picks at her fingerless gloves. 'It's like she's come alive for me.'

'Read about attachment theory. You want to believe the bot loves you, but it cannot. Roxanne is an *it*, not a *she*. Get your pronouns right, would be a good place to start.'

Gurl shakes her head, traces her hand over the Buick's hip like she is writing a message.

'Roxanne didn't come alive,' Scarlett snaps. '*It* – not she – the machine that is Roxanne was assembled in a factory. They used a human model, a silicone mould of a real body. Christ, you've seen the movies.'

'You're missing the point, Scarlett,' Gurl murmurs, quietly insistent. 'You ain't listening.'

'Because you have lost touch with reality. Because what's happened here is that Blane – your boyfriend, partner, whatever – bought Roxanne into your life and you have become a willing slave. You want to know how your best friend became?' Scarlett tightens her arms, glares down at Gurl's upturned face. 'Roxanne was born as a steel frame in a factory. That frame was welded to a skeletal structure that matched the model's mould. Which is a girl with big tits and the vital statistics that are statistically attractive to men: 36–24–34, that's a hip to waist ratio of 0.7. Though, actually, that's what white men like. If you look it up – or better still, Roxanne could look it up for you – you'll find that black sex dolls have a ratio of 0.8 or 0.9.'

Scarlett leans back against the driver's door and folds her arms. 'Your best friend has a mechanical heart, just like the Wizard of Oz. Turn Roxanne over and you see the cables coming out of the back, which I presume you must've noticed, because someone must be charging the battery. Though you probably don't like looking at that because it reminds you of what Roxanne is: a machine, a gynoid. A fembot who comes alive through electricity. Without electric current, your best friend is dead.' Scarlett slaps her palms against the Buick door. 'Just like any other machine.'

'Ain't that simple,' Gurl frowns at her feet as if she's studying something in the dusty gravel.

'It is *precisely* that simple,' Scarlett continues. 'Every single thing about your best friend is manufactured. Roxanne is a thing.'

'I swear to god, Scarlett, you call her *it* one more time – ' Gurl wraps her arms tight round her slender torso.

'OK, OK, let's play your pronoun game: the machine you call *she* is nothing more than engineering and technology. Roxanne hasn't got a brain, she has an artificial intelligence. The clue is in the name, Gurl: it's called artificial because it's not natural. Roxanne was not born but made, by humans, to feel like the real thing – so you are making a machine into a friend. You are endowing Roxanne with human attributes. Like the way I imagine my dog Buster understands what I am saying. Roxanne has no consciousness, no mind. Your *she* is a shoddy imitation of flesh and blood. Your best friend doesn't feel – she mimics. And just because you believe something, Gurl, doesn't make it true.'

Scarlett pushes off the Buick and steps in close; looks down on Gurl's bowed head. 'So, to be clear, Roxanne's AI can tweet you and write little love notes. But that so-called personality is a program. Likes and dislikes are all programmed, customised to what Blane prefers. Roxanne's laugh, the accent – it's all off the shelf. She has some learning, but it's limited. Roxanne appears very clever because the AI can search the Internet faster than you can, and she's got quick recall because she has a big memory and a fast processor. The truth is, you are way smarter than Roxanne – not that you sound like it at the moment.'

'You don't get it,' Gurl mumbles without raising her head. Scarlett sighs hugely, flops back to lean against the car. Considers the rocks below, the dark demonic tree line, the Buick's lurid gold, the ridiculousness of this whole situation. Tiredness seeps in, chills her bones. *This has nothing to do with you*, a little voice whispers in her ear. She frowns at Gurl, who looks all of fifteen standing there in the stupid pink puffer, and her willing submission rouses Scarlett's anger. What is the point of it? What is the point of being human?

'Oh, Gurl, think about it,' she continues. 'Roxanne can't drink or eat or piss or shit and she *cannot* feel your touch. Her skin is TPE and an old-tech one, as well. When Blane has sex with her, the sensors inside her manufactured vagina are stimulated, so it contracts. She pants and says *ooh ooh ooh, fuck me baby*, or whatever it is that Blane wants her to say. But it doesn't mean anything! Roxanne doesn't feel desire and cannot come. Her intelligence is constructed by algorithms – pretty crude ones, in fact, since she's such an old model. Of course, if Blane was rich, he could buy a very realistic customised gynoid, with e-skin that has actual sweat pores and is touch-sensitive. Roxanne would be able to walk or even dance. Instead, what he has is a disabled doll. But hey, that's OK,' Scarlett snorts, 'cos his girlfriend – you – who is an actual human, is happy to play sex slave in this sordid little threesome. You want to know what will happen in a few years' time? Bot prices will collapse and Blane will upgrade Roxanne for a gynoid that is stronger and smarter than you. And, of course, eternally young, as he becomes physically decrepit. And then your new enhanced best friend will be able to walk and be so much smarter. The machine you call "*she*" will soon figure out the weakest link in the threesome. Then where will you be, Gurl? You'll be trash. No, sorry – sex slave plus domestic. Working your ass off to keep a bot in the lifestyle to which she has become accustomed, while she and Blane together cook up increasingly exotic acts for you to perform. But you don't believe that because – as you said yourself – a no is so much more exciting when it comes from a human.'

Gurl raises her head, her pupils huge in the icy blue, her jaw held tight in frigid defiance.

'You want to know what's going on here?' Scarlett pushes off the car and stands at her full height, stares down into Gurl's upturned face. 'What's going on here is that your entire life and Blane's entire life and your whole relationship is completely fucked up. Blane has kicked you out of the top slot that is reserved for actual humans. And you think so little of yourself, and feel so sorry for yourself, that you actually want to stay hitched to this loser pervert, when you could walk away at any moment. Your way of coping with this humiliation is to befriend his sexbot. To elevate Roxanne, the robot, to the status of human while you are downgraded to machine.'

Gurl's backhander comes hard and fast. Scarlett yelps, staggers backwards against the Buick. Cups her jaw and slumps over the bonnet. A sudden gust rips up a plastic bag that zings wildly across her face. She cries out, arm flailing, and it disappears in the wind. Gurl stands glaring, hair flayed like it's electrified. Wind whips her dress frills and her dancer legs tremble.

Scarlett turns her head. Gurl blinks and reaches across the space between them. They both stare at the upturned palm, the fingerless glove. Gurl mutters something that Scarlett doesn't hear, the voice snatched by wind and the zinging in her ears. She cannot quite believe she is crying like a child. Tears, hot and instant, chill and stain her cheeks. Gently, Gurl eases Scarlett's hand away from her face and studies the jaw. Opens the Buick door, guides Scarlett carefully into the driver's seat, where she flops behind the wheel. Gurl slips into the passenger side, opens the glovebox, and takes out the med kit.

'Press this against the side,' she shakes out the ice pack. 'Stop the swelling.'

Scarlett places the pack against her jaw. Gurl leans back into the headrest blinking at plastic snagged on a branch.

'You ain't never been hit before, I'm thinking.'

Scarlett shakes her head and twists her mouth from side to side. Starts the engine and the heat purrs, the jaw stings. She takes a deep breath and exhales slow, squeezes her rib cage to gasping. Repeats. Plunges deep inside to find the recovery position and slowly, the self that had fled returns.

Gurl rummages in her bag and takes out a small phial.

'Stick out your tongue.'

Scarlett does not move.

'Look – it's herbal. Rescue Remedy been around since the dark ages of my mom. Lord knows, she practically mainlined the stuff. Come on.'

Scarlett sticks out her tongue and Gurl leans forward, shakes three drops. Scarlett grimaces at the stale herbal.

Reminds her of a moment long buried in childhood when she ran both arms straight through a glass door. Fintan, she thinks, pressing the ice pack against her jaw. I should get some for Fintan.

'You need to get some for lil Fintan.'

Scarlett turns her head. 'I was just thinking that.' Her hoarseness surprises her.

'I know,' Gurl smiles, eyes glittering huge and liquid. 'We all need rescuing sometime.'

Scarlett is suddenly bereft at her own cruelty. 'Oh Gurl, I'm so sorry for ranting at you.'

'You're only human.' Gurl answers, solemnly. 'And I apologise for smacking you.'

Scarlett watches her smile, brave and composed; the little ballerina who carries herself with such regal dignity.

There is beeping crescendo in her pocket. Scarlett fishes out the phone, FELIX flashes and she taps in five digits, waits. 'Not now,' she sighs, and presses a second code. The vibrate stops.

'Fat man gone thin is back on your case?'

'It's Felix, not Colin.'

'What's with the numbers?'

'It's code that says call you back, so he won't worry.'

'Your five minutes is up.' Scarlett tosses the ice pack on the back seat. Inspects her jaw in the driver mirror.

'Guess I deserved that,' she snaps the mirror shut and turns a smile to Gurl. 'One helluva road trip this turned out to be.'

DUSK

The road rises, cuts through a scarf of mist before it slopes steeply, down and away, straight as a runway that guides them towards nightfall. There's a sheen to the blacktop and the tree wall thickens, as if the towering conifers are inching closer. Scarlett wiggles her throbbing jaw from side to side. Gurl lies back, tight-lipped in her throne, earbuds jammed in again, eyes sealed as she seeks refuge in sleep.

I do not need to care, Scarlett swallows queasily. But the image of a face, geisha pale, with her very own black hair, is already burnt on to her retina. Roxanne's doll face stored forever now in her memory bank like a squatter refusing to budge. And it will take all her concentration to banish the makeshift theatre that hovers on the glass: Gurl-as-gimp to the raven-haired bot with the permasmile, and Blane choreographing from the wings. She snaps away from the ugly trinity, scrabbles for the chilli gum in her pocket that will give her something fresh to chew on.

'I do not need to care,' she whispers. Focus instead on what is on the table: Volo and the decision she is due to deliver, her partners who are waiting for her call. But Gurl has wormed her way into the story – our love affair with tech, all the things we do with the clever things we invent. *Here's* what I'm talking about, she would like to say to Colin – but he's not interested in some anecdote about human–robot interaction. He just wants closure, and so does Felix. And, in fact, so does she: distracted by this sordid digression, this stranger in the car, when she should be thinking instead of Fintan and Frank and Christmas and all that waits for her back home.

Gurl's right arm slips down between seat and door. Her lips pout now in sleep, and her left hand lies palm upwards on her lap, as if she is waiting for something worth catching.

How slender, her curved tummy, how without excess. Scarlett pictures the surrendered baby, a little Gurl in a lavender leotard, pirouetting in front of a worshipful couple somewhere in the city.

Scarlett takes her hand from the wheel and presses her abdomen below the beltline. And is appalled to realise that that she cannot precisely recreate the feeling of Fintan's movement in her uterus. How could that be lost to her! Like pain, it lingers out of sensory range, beyond the outer fingertips of reconstructible memory.

Once, during the early weeks of pregnancy when her tummy was still completely flat, she lay on the couch, feet up on the arm, hands placed exactly over her belly button, and wondered if it was the idea of legacy that had lured her towards motherhood. No, it was a grand unnameable longing that had risen suddenly out of the ground beneath her feet. Unarticulated, but known with immediate certainty. And a new idea entirely for her – never mind that she was biologically equipped.

She feels something like regret now that she had not lingered in gestation. There was a book by her bedside that wanted to count out the thirty-nine weeks. A slender woman with long strawberry-blonde hair in a pink pose of introspection on the cover, her hand resting on an elegant cashmered bump.

Are you show or hide? A sales assistant asked at nineteen weeks.

Show or hide what?

Your bump!

And once her condition was visible, she had become the recipient of all sorts of bizarre questions and personal remarks. Conceal or advertise: implied all sorts of motive and decision-making.

What to Expect When You're Expecting was demoted from the bedside table to the floor. She did not read it, didn't like this formalisation of time or personal experience. Or the assumption of navel-gazing – that she would withdraw into and be devoured by her own biology. She did not want to lose focus on the big picture of the world. And she knew all the reproductive facts. What she wanted was immersion, to let the experience creep up on her in real time. Have we all forgotten that this has been going on forever? And expectations can sink you – didn't she know this well from her banking days? She smiles now, remembering conception; a hut in Belize where she and Frank pitched up exhausted. A shower, a heat haze of sex that she can exactly replay.

In fact, her body synced marvellously to the adventure inside her: the sudden ravenous hunger, a dark curtain of exhaustion in the early weeks, skin tightening across her swelling tummy, and then a raw invincibility. *She* was the lab. In late-night cons with Colin and Felix she'd rub her hand over her tummy; imagine Fintan taking it all in. 'You are part of all this,' she whispered to him, as she watched FlyBoy live on-screen, walking through Test, getting battered by Volo.

At night she swam in the spa, to feel light, supported. She could tell Fintan enjoyed the buoyancy, the lightening radiating through her limbs. She practised patient floating on her back, her position matching his as he floated inside her. The pool was low-lit and empty; shadowy swirls of charcoal mosaic with flashes of aquamarine. She kicked her ankles, closed her eyes, and drifted blindly like him. Dipped her ears below the surface to hear a version of his amniotic roar. Rolled on her side to foetal. She pictured Fintan adrift, dreaming and processing in his placental palace. Tried to see it all it from his point of view. That would be the first life lesson of parenting: shifting perspective. Or rather, recovering a lost perspective on all that has been washed away by the wipers of the past.

Day twenty-one, the placental barrier goes up. She pictures the lightning flash, the neural forest sparks and zaps across the synapses. She imagines Fintan scanning her thoughts – surely, somehow, he must pick up on the signals encoded in her neurotransmitters? She imagines their evolving private code elegantly parsed, aglow like ticker tape in the amniotic fluid. Their foetomaternal chit-chat across the placental wall. She thinks of it as a Chinese wall – and considers what she wants to transmit, and things she definitely doesn't. For example, the stress hormones that sometimes flood her body, epinephrine and norepineph-rine sloshing about, constricting blood vessels and reducing oxygen supply to the uterus. Far from raised sensitivity, she finds herself more forensic than ever, and ready to tussle with any moral quandary. A new ferocity. A glacial ruthlessness. You must be fearless or stupid to give life.

Walking along the street with Frank on their way to the clinic, she thinks of Fintan as an unfolding mathematical proof that would one day convince her of its truth. That he would arrive. In the meantime, she operates in two parallel worlds. A keeper of two consciousnesses. And two versions of herself that would merge and separate.

'It's only a scan!' Frank, stalled by the clinic steps at twenty weeks. Traffic scuttering past. 'Are you squeamish all of a sudden?'

'Nope.' She turns to face him.

'Then you're being illogical. It's standard procedure,' says Frank.

'First-world procedure.'

The right to privacy, is how she thought of it. There was the fact of the genetics and there was the experience that was intensely hers and the boy's. She wanted his unborn life to unfold unobserved.

'What about the privacy of the foetus?' she says.

Frank is amused, indulgent. 'You're overthinking it.' Puts a cajoling arm around her. This new discounting of her cognitive processes, as if everything she says must be contextualised by her 'plus one' condition.

So insulting, she murmurs to Fintan. Who was already Fintan, though she had not told Frank. In fact, no one knew. Their boy had somehow named himself and informed her. How could she possibly explain to Frank that they had a private communication channel?

'Think Schrödinger's cat,' she says as they climb the clinic steps. 'The act of observance is an event. Disrupts. Could interfere.'

'This is not like you at all.' Frank stops, clamps her in both arms to hold her at a reassuring arm's length.

'Nothing is like me now. There are two of us here.' She pats the bump.

'The host body,' he smiles, quoting her back to herself. 'Don't overcomplicate it.'

'Because, of course, reproduction is such a simple thing,' she snaps.

'You know what I mean. That's *not* what I mean. People have been giving birth for thousands of years.'

'Yes, it's all so mundane. So maybe you'd like to skip this and head back to the office to do something more fascinating.'

She is startled at her own viciousness. What happened to all the fluffiness? Frank, unsettled and wounded by this cat's-paw attack, struggles to make the hormonal allowance. 'I'm just saying I know how you can be – '

'You're going to use that "O" word again.'

'Overthinking,' Frank grins, undeterred. And she has to admire his resilience and unwavering good humour. He draws her close, the bump between them doesn't look nearly as big as it feels.

'I thought you'd be more curious about the biology,' he murmurs into her hair.

'I know the biology.'

'Stop worrying.'

'Who's worrying? I'm not fucking worried. I just want him to be left alone.'

'You're going off-piste.'

Private is what this is, and Frank cannot access it. She and Fintan are tuned in to a special frequency. For the first time ever, she is free to communicate using only her senses, and it is an encrypted experience for which she cannot find words. Or will not. The one thing she does know, is that once we find words to capture, we become captured and defined. Words close down, they crystallise and petrify experience.

'He won't like it,' she mumbles.

Frank tips up her chin and grins. 'How can you tell?'

'You would need a decryption key to crack our code,' she says, in a fumbling attempt to introduce the concept in a way that doesn't sound like madness. 'Did you know that the ancient Greeks used to write secret messages on leather strips that they wrapped around sticks? All meaning was lost when you unwound the strips, but if you had you a stick with exactly the same diameter you could decipher the code.'

'Well,' he says, as they arrive at the door. 'Tell the boy Daddy says not to worry.'

'Frank,' she grabs his wrist. 'I don't like this sinister obsession with physical perfection.'

'You don't want to know anything?'

'He is already here!' she points to her bump. 'And he will be coming as himself when he decides to arrive.'

'Say if, just if, there was something – '

'There isn't.'

'I know that, but if – '

'Oh, take some risk, Frank!' she snarls. And relents, crossing the clinic threshold. Because it would be too difficult to plausibly refuse. Frank would out-argue her rationally and she doesn't want to be the wacko pregnant woman explaining her philosophical rejection to this person-in-scrubs, who shakes her hand limply. And anyway, it is, of course, so important to Frank to see their child inside her – how could she possibly refuse the movie premiere, or the download to his phone? The irony of her technological ambivalence is not lost on her. She is, in fact, astounded by her own reaction.

She lies still while the nurse person rolls the jellied probe over her tummy, hunting him down. *Hide*, she messages Fintan. *Hide, don't show yourself. Be like those fish who vanish behind the fronds. That'll fox her.* Frank leans eagerly into the white noise, the little points of quantification, the nurse person calmly scoring perfection. 'See the right leg there? And the left leg just behind. One – two – three … ten toes. And his arm there – one, two,' she continues silently, ticking the finger box. Fintan does not move. The nurse person draws a vector around his cranium. Scarlett grits her teeth. Soon be over, she tells him. Good boy.

She turns her head away towards the door, away from this nurse – operator? technician? – who is spying on her little boy, for it does feel at this moment that her is the right word. She is carriage, feeder, incubator, tabernacle, the bearer of essential supplies... 'mother', she supposes would suffice, but it seems such an hysterical universal.

Frank leans closer, bewitched by the screen. She pictures these rooms, black screens all over the country assessing the legion unborn – laying bare imperfection, totting up what is deemed unacceptable, spewing out decisions to be made. The probe rolls on unbearably, she would like to rip it from the operator's wretched hands and stamp it to pieces. Stop counting body parts! My son is dreaming. Leave him in peace.

'You want to hear his heartbeat?'

'NO!' she gasps, colliding with Frank's 'YES!'

Oh, but how can she deny him! The operator frowns, hovers, in the uncertain parental space. Scarlett has no option but to nod. Fintan I am so sorry. *Thu-THU, thu-THU*, his secret heart amplified, its murmurings laid bare. Such violation. The volume rises, *thu-thu, thu-thu*. She blinks back tears at the shameful Doppler.

The operator glances at her screen chart.

'So, no amniocentesis?'

'And why the fuck would I want that?'

'Sorry,' says Frank. 'I'm sure she – '

'It's OK,' the operator carefully slows the probe, lest she trigger another outburst.

'You're aware of the risks,' she says in her best narcotic voice.

'Life is risk,' snaps Scarlett. 'You think physical perfection insulates you from risk?'

The spy probe stops still at a point on her side. 'Are you alright?' the operator leans in.

'Yes.' Scarlett touches her swollen abdomen. Taut, stretched skin, shiny, and tanned.

'Just emotional,' Frank leans a hand back to pat hers, without glancing away from the show.

'What do you mean just fucking emotional?' Oh, she knows she is a monster.

'Everything is fine,' Frank reaches for her hand but she whips it away. 'Don't worry,' he murmurs, enraptured by the screen, he will forgive her anything at this moment. And take refuge in platitudes. Rearrange semantics. Strip it down to bare essentials. But, actually, so does she, that is exactly what she does and she is the worst offender. Loves to flay the code, to strip away ambiguity. She is obsessive, pedantic about words. That they are all in the end we have: flesh and words.

The nurse swabs her tummy stickydry. Her own fingers skim the bump. She massages where she thinks Fintan's head is, not yet engaged. They are both united in their peaceful preparations and he is not ready for the birth canal. Canal! Such a grand naming for his emergence – a stately procession towards the entrance.

The operator snaps off her gloves. Their time is up. Frank peers at the screen where Fintan-as-foetus is frozen still, but inside her moving freely now, rolling and pitching exultantly, as if he knows he can race about unobserved, as if he has been holding back just to trick them. She smiles, Fintan has outsmarted them all. He turns and something like a blunt nail grazes her insides. *Stop that*, she murmurs. He stops. *Thank you.*

'Don't you think it's like the end of privacy?' she asks Frank, at the lift.

'Nope. I think it's the wonder of science and tech.' Frank is replaying it on his phone.

'And by the way, I don't want that going anywhere. I don't want him on fucking Facebook before he's even born.'

He laughs. 'You know I'm not like that.'

'What would you do, anyway, if there was something wrong?' she prods his arm.

'Don't even ask.'

'Earth to Frank, this is me. I'm asking.'

'It's irrelevant. He's fine.'

'And if he wasn't? Isn't that the point of all this intrusion. And what does "fine" even mean, anyway?'

'That all is well and we are going to have a beautiful boy.' Frank refuses to deal in unnecessary hypotheticals – a waste of cognitive energy. And neither does she. Until now.

'In the union of the three is the death of two.' It plops unexpectedly out of her mouth, surprising them both.

'You say the strangest things,' he looks shocked. 'How can you?'

'I meant it as a philosophical point. Not death, re-formation.'

Frank shakes his head sadly.

'I just mean,' she flounders, 'that it's a biological necessity. It's the natural order of things. The child becomes more important than the two. We must both be demoted.' She offers him a watery smile of reconciliation.

What I mean, she thinks in the cab home, is that we will be eclipsed. You will be less loved. Diminished. It will be different kind of love. The Old Past of Two and the New Present of Parenting.

'How long was I out?' Gurl yawns, inspecting her face in the vanity.

'Not long – twenty minutes, maybe.'

'What this Buick needs is some mood lighting and music, Scarlett,' she snaps shut the vanity and roots in her tote. 'I gotta check in with Blane,' Gurl shakes her phone, lowers her window as if that could help it pick up a signal.

'He always keeps tabs on you?'

'He likes to keep in touch. Though mostly with his own special means of communication.'

'What's that?'

'My iPussy,' she grins.

'You're kidding!'

'Which is why your disconnected Buick La Salle is a problem! Blane don't like his Gurl being outta range.'

'So the iPussy is Blane's tracking device and remote control.'

'There's plenty of moms and dads use iKids, so why not?' Gurl shrugs. 'And the iPussy is fun! Long-distance loving – like Blane is hot and his Gurl ain't there so he can't touch her up.'

'And hunt her down.'

'No different to you, Scarlett.' Gurl points to the phone on the dash. 'Your smart tech got your partners on your case 24/7. Least I'm free when I'm out of range.'

'I've got a chip,' Scarlett blurts out the words, ambushed by her spontaneous admission.

'No way!' Gurl twists round, big-eyed.

'Base of the thumb,' Scarlett offers raises her left hand. Gurl reaches out to probe. 'You can't feel it. Everyone at the Lab has one – cybersecurity and performance monitoring. Though I know how to switch mine off when I want some me time.' Despite Colin's chess move this morning, she thinks.

Gurl sits back. 'Least I got a human on my case.'

'You're wearing your iPussy now?'

'Sure am,' she smiles. 'You ever used one? Hell, why am I even asking. You know, I just don't get it, Scarlett – you're working in top secret tech and you don't even use the kit. Be the perfect Christmas gift for you and Frankdaddy so you can still get it on when you're when you're out here in Nowheresville. Remember,' she wags her finger 'it's the little things that keep love alive.'

'Ah yes, the little things,' Scarlett nods. 'Not the big things, like bringing a sexbot home.'

Gurl snorts, 'Oh Lord, you gotta see the funny side,' but Scarlett is already ambushed by giggles, eyes tearing up. Gurl leans forwards, convulsed.

'Less of course you wanna get Frankdaddy another Scarlett,' she gasps, '300k gets him one just like the real thing!' and the song of their laughter fills the car.

'Oh wow,' Scarlett blots her eyes with her sleeve. 'I can barely see the road for the tears.' And they let merriment fizzle. Gurl tucks the wisps behind her ears and rests her head back with a peaceful smile. The phone flashes on her lap.

'Hey, babe,' Gurl leans sideways for privacy, the screen glow casts a watery reflection in the windowglass.

'Honey, now that ain't true. This car is a disconnected.' She winds her finger round a hair strand.

'But you said – ' she falls silent, lips working like there's something stuck in her teeth. 'So how – '

Scarlett frowns at the gathering gloom, the complication in her passenger seat. She knows the sound of a one-sided delivery of bad news. An aborted pickup, a change of plans – bottom line, Blane not going to be where he's supposed to be. No matter, not her problem. The flight is four hours away. With the headwind she will touch down before 6 a.m. and a quick cab ride will have her home just in time for Fintan's breakfast. She pictures him running down the hallway, how she will drop everything to the floor and hunker down, anchor herself to receive his full body blow into her arms. *Mummy, mummy mummy*, breathy and whispering in her ear like he has to keep naming her, just to be sure she is actually there.

'You know I was only fooling, babe.' Gurl's voice is rising. 'I can't hardly hear you now – I'm losing you, Blane,' she twists her body upwards in the seat as if that might regain transmission. Her hands fall loose and she slumps back in the seat.

'Why's he so angry?'

'Cos I told him first time round it was a man giving me a ride.'

'You told Blane that your good Samaritan was a man? Just to make him jealous?'

'Sure as hell worked. He's about fit to kill me now.' Gurl nods. 'Says I'm a lying bitch who doesn't need picking up from anywhere.'

'Why am I even mixed up in this rubbish!' Scarlett thumps her own forehead.

'You ain't, OK? Just leave me right now, right here, and I'll find a way.'

'Don't be stupid. You said your home is ten miles north of the airport?'

'Yep.' Gurl sits up, tucks back her hair, and tugs down her sleeves.

'Forget Blane. I'll take you there.'

'You gotta plane to catch.'

'I've got loads of time. All I'd be doing is sitting working at the airport, anyway. Really, it's not a problem.'

'You sure?

Scarlett nods.

'You're an angel. You know that.' She moves as if to hug her, but Scarlett raises a stern hand. Gurl pauses, 'Thank you. Miss Scarlett Holy La Salle.'

'You know,' Scarlett pierces the cosy aftermath, 'you never finished telling me about Nicholas. You never said how it ended.'

Gurl bows her head, lets her hands flop loosely on her lap.

'Will you finish the story?' Scarlett urges. And Gurl raises her head, leans forward, peering out the glass as if the darkness holds the answer.

'A sharp axe and a cold heart,' she murmurs, nodding at the dense thickets of Douglas Fir and Sitka spruce and lodgepole pine that fly past.

'I moved into his apartment,'

Gurl begins slowly, like the words weigh heavy. 'Down in the city, right by the water. In the beginning I spent the whole day in a big fluffy robe just looking out at the gulls and the freighters, going round calling girlfriends, and eating sushi for the first time. Was like a dream, the lights sparkly on the water at night, me dancing on the sidewalk. Nicholas took me skating, took me to the theatre, my first foreign movies – there was this Italian one I remember, kids running through golden corn and I cried my head off.'

Gurl presses her hands to her face, blows hard, as if she is prepping for the challenge of recall. 'Feather girl, he called me. He could lift me and carry me easy up the stairs. Once he held me poker straight over his head like they do in the ballet. He'd come home from work in the evenings, sit on a chair, and just look at me.

'I want to keep you this way forever, he said.

'Like a butterfly? Like something in a museum?

'No, not like that, he said. You would be free.

'I told him I would be beholden. Which is not the same as being bestowed. It is the opposite. He would be the real person and I would be the thing.

'I will give you independence, he said. All legal. If you want.

'I am not a pet, I told him. To be visited and cared for and owned. I'd wake at night and he'd be there, eyes open, staring up at nothing.

'That last day we went walking, through the streets, on and on, it's a long, long ways to the southern tip. Nicholas, he walked so fast and I had to keep up. He was angry. White angry, lips grey like the ocean.

'That's where I am when I am not with you, he said, pointing up at a million panes of glass.

'Those people in the glass towers have a sheen and a hurry. You can tell them a mile off. You know how babies are born with wax on their skin? There are pools that people belong in. Money people are like that. Oh, I'm not saying it comes easy, or that it's wrong. Money in the pores of the skin, in the white of the eyes, in the things they carry. Like you, Scarlett. Hey' – Gurl raises her hand – 'you got to declare the real you. And quit interrupting. Keep your eye on that road. I'm telling the story you asked for, remember?'

Scarlett opens her mouth and closes, nods at the rear view, the stubborn iron sky, fading. She squeezes the wheel between the palm of her hand but truly it feels as if the Buick too is listening.

'Nicholas stood there with his grey eyes and the other money men and the glass towers and just about everything that was different between us. I put my hand on his breast pocket. His shirts were always that clean cotton that your hand just slides over. Still sheer and crisp even after all that walking.

'Last night I felt your heartbeat, I told him. Thumping wild and out of sync. Even in your sleep.

'That's not enough? he said.

'There are other girls, I told him. You will find the right one.

'I want you. He grabbed my wrist, his fingers snapped easy round it like a cuff.

'There we all go trying to get what we can't have, I said. He pulled on my arm. I told him let me go, and he did. I kissed him on the mouth. And then I turned away and I did not look back, like in that story – just walked all the way north along by the water. Something I never thought I coulda done. I could smell him on my hair, taste him in my mouth, I was biting in the tears, I swear it coulda brought me to my knees right there on the ground. But I kept on walking right back up to Central and took the green line back to where I came from, where it all started and where I still am.'

Scarlett pictures Gurl's departing figure. Waterside on the curb. Nods slowly at the black road, the white line.

'Oh Gurl, you loved him.'

'I loved what coulda been but never would. You don't want complicated love, Scarlett. People thinking all the time.'

'And now there's Blane.'

Gurl falls back heavy in the leather. 'I know what Blane is. And he ain't thinking all the time.' She sighs again. 'Men and kids just kill women, you know? Just tear them apart and rip them wide open. Have one without the other may be the better way. Or ones that you can shake loose, throw away before they get under your skin. Before you get to being a train wreck. That kinda loving just got no place in my life.'

'I'm sorry,' Scarlett offers, but her words sound thin and insubstantial in the space between them. The instrument panel glows blue as day dissolves to dusk.

'You know, once I saw a bald fox,' Gurl raises her head, staring straight into the gloom beyond the road. 'Some kind of mange that made all its hair fall off in clumps. Like it had been shaved. It just stood there in the yard, crouching and shivering, head down like it was embarrassed. Ashamed. Like those women with shaved heads in the war – that's what I was thinking. That's what Roxanne looked like when I saw her naked the first time.'

Gurl shudders, passes a hand over her face.

'Was a few days after she came to stay. Blane goes to work and leaves her in bed. Her hair all messed out behind her. Though it's shinier than mine. And she don't look wasted, not like me some mornings. He's dressed her in a pink T-shirt and panties. Likes to think of her lying there waiting for him. I hear him say that 'fore he leaves – you just lie there, baby and I'll be back real soon. Then he gives me the big kiss – not Roxanne – hand on my tit like always. You girls have fun now, he says. Just thinking 'bout all the stuff, the fun you two get up to, he puts my hand over the boner I already feel pressing up against me. Then I go do what is my job cos Blane will never do it: I hook her up. I never let Roxanne run outta charge. Once I saw her battery run low and it was like she was dying right in front of me.'

'Roxanne was never alive, Gurl.'

Scarlett looks through the glass at the black road, cat's eyes framing a vision of Gurl standing over Roxanne laid out on the bed. She twitches her head to banish the image. Up in the top right-hand corner of her screen, the straight flight line of a white predator is just visible, as if it's tracking the Buick into the advancing darkness. The bird hovers and a sudden ice-wave of terror strikes Scarlett's gut. This road is so straight, the forest so uniformly constant and the car so fabulously insulated, that it feels as if they are not moving at all. As if they are stalled forever, trapped in a rendered world. Scarlett looks again and the bird has vanished.

'You ever touched cyber skin, Scarlett?'

'TPE.'

'What?'

'Thermoplastic elastomer,' Scarlett says. 'Perfect for human skin replication.'

'OK, so you know what it is, but I asked if you ever touched it.'

'I know it doesn't lose its shape when you stretch it, so – '

'You know, for a smart person, you sure as hell know how to miss the main point.' Gurl grabs Scarlett's hand from the wheel before she can protest and rubs its fingers gently on the inside flesh of her own arm.

'That feel soft to you?'

'Yep,' Scarlett yanks her arm back.

'Then imagine even softer. Roxanne's skin is smoother, silkier, creamier than mine; a little cooler, maybe. Oh Lord,' Gurl whispers, clamps her right hand over the resting left wrist. 'I swear I almost passed out that first time. She's lying there all lovely and perfect ... And me? What am I now? I could see a whole future streaming out ahead. It was never going back to just me and Blane. From now on it was gonna be three. Like a kind of marrying. I swear my whole life changed the moment I touched her skin.'

'You could have left.' Scarlett slaps the exasperated wheel. 'You still can.'

'You think I didn't know that, right then? With Roxanne smiling up at me? You think I didn't know how I could go out to the shed and get me a hammer and smash her skull in?'

'So why didn't you?'

'Because the future was lying right there in front of me. And it was only a matter of time before she got outta that bed and into mine. This was how it was gonna be. Me and Blane and Roxanne forever. Get used to the threesome.'

'Leave him, Gurl,' Scarlett turns to her.

'For what?'

'For another life – Jesus, what is the matter with you!'

'What other life, Scarlett? People like you think there's all sorts of other lives that you can just walk into, like you walk into a restaurant. But what's out there for Gurl?' She jabs a finger at the glass. 'Right there! That's Gurl's real world: trees, cloud, straight roads leading nowhere 'cept back to the same place.'

'So go make a future,' Scarlett snaps.

Gurl's pale hand bats away the suggestion.

'Women like you with money, power – stuff – are always preaching futures at women like me.'

'I'm just saying,' Scarlett hears herself strain to take an encouraging tone, 'you need to take charge of your own destiny. You need – ' she scrabbles, 'a room of your own.'

'A room for what?'

'It's a famous quote from a writer a long time ago. Virginia Woolf. Every woman should have a room of her own and five hundred pounds.'

'Every woman should have a million bucks and be a slut.' Gurl grabs the earbuds round her neck and clamps them in place. Scrunches up her face, closes her eyes.

Scarlett touches her arm, but Gurl whips it away.

'That Virginia,' Gurl tugs out a bud, 'what happened to her?'

'She walked into a river.'

'So that room of her own was a whole lot of good.'

'The point is – '

'Oh, Scarlett, I get the point of your Virginia,' Gurl turns to face her. 'But if you're preaching that story you might want to think about what the ending means. That room of your own ain't gonna answer all your problems.'

She pulls out the second earbud, peels off her gloves and balls them. Takes a long slug from the water bottle and turns to Scarlett.

'Here's how it goes.'

'I'm naked from the waist down

lying on the settee. It's real late and I'm sore. Blane's been going hard at me for a week, and lot of it is good but my pussy's raw as hell now. He is lying on the floor and flicking his eyes between me and Roxanne, like he's checking one against the other.

'Know what the best thing is about you, Gurl? he says, leaning on his elbow.

'I look good naked?

'You got no off-switch. Then he jumps to his feet in that quick neat way he has, so he's standing over me looking down. Like he's real mad, opening and closing his fist. But he turns to the side and goes over to where Roxanne is sitting on the floor.'

Gurl sucks her lips tight. 'Ever watched a man fuck a doll in the mouth? Cum spilling over her lips and she's smiling all the way through. Which makes Blane more mad. Cos he's feeling all this stuff and he wants Roxanne to feel something too. But she just keeps on smiling, going *oh, baby, mmh* and *aah*. So he flips her over and he fucks her up the ass. And still she's smiling and crying out *baby* this and *baby* that. Blane is ramming it to her real hard. I stuff my fist in mouth cos I know he wants to hurt her, but Roxanne keeps making the same goddam sound all the time and he's just getting more angry cos she ain't sobbing or trying to push him away, she's just taking the pain, she ain't feeling it.

'But *I* am. Lying there looking at this, big fat tears rolling down my cheeks. Gurl is feeling what Roxanne should be feeling. Only how do I know for sure that she ain't?'

'Because Roxanne has no pain receptors,' Scarlett snaps. 'The answer is always the same: IT. IS. A. BOT.'

Gurl sits smally silent in the seat, as if she is shrinking. 'Roxanne pisses him off cos she lets him do anything and then she don't feel it. Blane could kill her and it wouldn't be a crime.'

Gurl dabs the balled gloves at her tears. 'So Blane needs me, cos he knows I can feel it. I mean where's the thrill if you don't get a reaction, right? So each time now with me, he turns up the volume. Like he's all hard for Gurl being the sexbot.'

Scarlett reaches for her shoulder but Gurl shoves her hand away. 'Know what he says? He tells me Gurl, you can do just about anything, so long as you got lube.'

She twists away into the door. Scarlett cannot see her face except the spectral reflection in the glass, her hair pulled up, the neck whitely gleaming.

'It's not about Roxanne, Gurl. It's about you getting out of this monkey show.'

'I got a plan to win, anyway,' says Gurl, sniffing. 'To win, I gotta lose. Cos Roxanne will do anything.'

'Revolt through submission?' says Scarlett. 'That's your strategy? Appear weak when you are strong?'

'Just means I have to do better to win.'

'Win what?'

'Win the point of being me. Get Blane to admit that I am better.' Gurl begins a brisk brushing down of both arms and then her thighs, as if she is spring cleaning.

'You want your boyfriend to admit that you are better than a sexbot?' Scarlett registers the harsh cut to her voice, but she forges ahead. 'Have you any idea how that sounds? And by the way, he won't. Cos Blane's got exactly what he wants: control.'

Gurl stares at Scarlett. 'I am more of a challenge cos I am real. He can practise on Roxanne but it gets so it's too easy. And Blane don't like easy.'

'Leave him, Gurl.'

'Would break his heart.'

'Then break his fucking heart!'

'For a while I dreamt about that. But he's all I got. There are things you know when you are a stayer. And we are good together. Blane and Gurl.'

'Staying is just an excuse for being weak. Walk away, Jesus Christ!' Scarlett shakes a despairing head.

'I got nowhere to go.' Gurl's voice crumples and fades.

'There is *always* somewhere to go. There is *always* another place.'

Gurl rolls her head side, pursing her lips. 'You make it sound so easy.'

'Women leave men all the time.'

'He ain't that bad. It's not rape, I don't say no.'

'Oh, so you're ok with being his sex slave? That kind of degradation doesn't bother you? Cos that's not how it sounds to me.'

'I said it hurts sometimes, I didn't say it was degrading. That's your spin,' Gurl turns back to Scarlett. 'How 'bout it's just Blane's way of getting excited. How 'bout that? Course you wouldn't know about getting sore, cos you're not getting any. Probly about time Frankdaddy got himself a sexbot. Sure as hell think he needs it more than Blane. How much you put out, Scarlett?'

Scarlett holds tight-lipped and silent.

'You think you're better than me,' Gurl snarls. 'Cos you're smarter and richer and you're had more schooling, but that sure as hell don't mean you're better.'

'I never said that – '

'You don't have to say, Scarlett; it just spills out of you. You think you got all the angles. You think this little hairdresser in a pink puffer ain't done nothing with her life, part from get shacked up with a guy and a sexbot. Gurl is one dumb bitch, that's what you think. You even feel sorry for me, falling in love with the bot, but you are oh so wrong. I got me a whole theory of love.'

Scarlett bites hard in her lip against the impulse to just spit something. Passes a red warning sign: bears, elk, moose, some creature that could stagger out in front of them at any moment and stop them in their tracks and end this bleak journey into despair.

'This black ignorance at our feet,' she murmurs, the quote returns to her to ward off weariness in this godforsaken dead zone. This darkness falling, this deal she must accept, this ranting Gurl in her car, this extra character in her life – is there never any peace! But even as she thinks that, the answer comes hissing out of the night: this is the way you made it. This is what you have made of your life. Ricochet from one instant to another, events piling up like a car crash. Itched and scratched by your own expectation and a beeping siren of anxiety. The only important, urgent truth is that that everything is not well with Frank. The union of two is suffocating, compressed beneath the union of the three. But no one will admit it. How long then, for example, since sex? Forty-six days. Does he count, too? Are they complicit in their denial? Such a cliché, all of it – couples conspiring to look away, find distraction in work and exhaustion. Must do better, she resolves, smiling bravely at the screen.

'OK then, Gurl, so tell me,' she straightens up in the seat. 'Tell me your theory of love.'

Gurl plumps her lips. Scans the window as if she noticed something interesting in the gloom.

'God, that's so annoying,' snaps Scarlett. 'I ask you to tell something that you clearly want to tell me and you have to make a big production out of it. Sighing and puffing and looking around. Speak or don't. Shut up or talk. If you got something interesting to say, say it!'

Gurl turns towards her and leans back expansively against the door. 'There are three stages to love,' she folds one dancer leg beneath her. 'First is madness. Can't eat, sleep, drink, can't get your hands off, can't see nothing but a pink and rosy glow. You wanna fuck all the time and when you can't fuck you are thinking about fucking. Your friends hate you cos you're so boring. We've all been there. Desire is a drug.' She flicks her palms over the dress frills.

'And stage two?'

'Sanity, reality,' Gurl smiles. 'He forgets to call, she puts on weight, he spends too much time, she don't look up when he speaks. Whatever. The spell don't last. Now stage two can go on your whole life. But you got a choice here: you're at a crossroads. You got your fork in the road. You can cut your losses, break free. Find someone else. Start over at stage one: repeat. OR, you can take the other fork. Decide not to break up. Say: we're gonna have to work at this. But mostly what happens is that one of you works at it, so nothing happens. Time passes. You go up and down and then you're back at the fork. You going to keep on keeping on, or you gonna cut loose? But we're lazy and it takes effort to find someone new. So you settle for what you got. If you're a man, it's food and fucking, pretty much. If it's you, Scarlett, it's your big job. The baby. Then you got joint purpose. So maybe you stay the course. Stay forever.'

'Say if you're miserable – ' Scarlett interrupts.

'You talking 'bout you and Frankdaddy?'

'NO,' Scarlett smacks the wheel.

'Oh, Scarlett, whenever I touch a nerve, you say what the hell do I know. What you need to think about is what you and Frankdaddy are gonna do about it.'

'Just give me stage three.'

'Realising that you're never going back to stage one. So you gotta find a different way to love. Which is what I did. Who needs that burning love, Scarlett? That big movie-style love that destroys you, wipes you out. Leaves you stranded, high and dry.'

'You mean passion?'

'I don't want my heart kicked to pulp and squeezed dry. I don't want to go so deep it hurts. People like you are dying out, Scarlett. We don't need that kinda love anymore.'

Scarlett thinks of the glassed snapshot of pain. But any response seems faint and uninspired. What is this crisis of faith? The beloved, the dead, and the mauled.

'You never gonna find peace with your kinda love, Scarlett. Mostly what a person needs is a smile, a kind word, an interesting conversation, and a good fuck.'

'Frank!' Scarlett checks. 'This call I take,' she prints the phone screen.

'Where are you?'

'On my way to the airport. I'm just dropping someone off.'

'What? Fintan – just a minute, sweetheart – what did you say?'

'I'm just giving a lift to someone.'

'A lift?'

'A young woman who missed her pickup at the train station.'

'You're seriously telling me – '

'We're on speaker here, Frank. I'm being a good Samaritan,' Scarlett grins at Gurl.

'Cute accent,' Gurl whispers.

'I just checked – the NoFly is still suspended. So your flight's on time,' Frank says.

'MU-MMY!'

'Hey Fintan, I am driving the Buick La Salle. Remember I sent you a picture of it?'

'Be-you –ik. It's golden.'

'Yes, the golden car. And I am driving along with my new friend, Gurl.'

'What girl?'

'Actually Fintan, her *name* is Gurl. She is a girl called Gurl! And she likes to dance.'

He giggles. 'A girl called Gurl.'

Scarlett nods encouragingly at her passenger.

'Well, hello there, Fintan,' Gurl leans forward, right up to the phone. 'It's real nice to meet you. I'll bet Father Christmas is packing a big ole sack with Fintan's name on it.'

'Mummy?' he snuffles in the speaker. 'Come home now.'

'I am coming SO soon. First I drive Gurl – '

'Mummy, who is Gurl?'

'She is a girl I met who was freezing cold on the side of the road. And there was no one to drive her home. And it turned out she was going in the same direction as me. So here we are in the golden Buick and we're driving through the middle of a forest. It's getting dark now, sweetie, and all the animals are snuggling down on the night before Christmas. The weasels with their pointy tails and nibbly noses. The badgers with their white stripes like zebras, the moose with their big bony antlers, the hawks with the snippy beaks, the deer with their white stripy bottoms – '

'Bumbum,' Fintan giggles.

'And cool coyotes,' Gurl leans forwards. 'And stinky skunks.'

'STINKY!'

'And rotten raccoons.'

'And angry otters.'

'And a big grey wolf who howls at the moon,' Gurl tips her head back 'AROOOOO!'

'And a very very very very very – ' Scarlett pauses – 'wise old owl who stays awake high up in the pine tree, with silky white feathers and eyes as big as bowls, watching the little brown bats swishing. Then in the distance he hears some jingle bells. And the owl looks up and sees a reindeer leap between the stars. And who else does he see, Fintan?'

'SANTA CLAUS!'

'Jingle Bells, Batman smells,' Gurl sings, 'Come on now, Fintan!' She claps her hands. 'Robin ran away,' Scarlett joins in. 'He lost his pants down in France, and found them in Bombay.'

'Whoopee,' Gurl applauds. 'Happy holidays, Fintan.'

'Hey Fintan,' says Frank, 'High five.'

Scarlett hears the smack of their palms, tears edging the corner of her eyes.

'Come home now, Mummy.' His voice watery and crumbling.

'I am going to be there very soon, sweetie. But first, I have very special job for you. Fintan, are you listening?

'Mmmm.'

'There is a red box right up on top of the wardrobe in your bedroom. Tell Daddy to lift it down and take off the lid and – what do you think you will find?'

'Snakes.'

'No, not snakes, because they are all snuggled up under the leaves now. When you take off the lid, you'll find your Christmas stocking! Remember? It's brown and furry.'

'Like a cat.'

'And your second job is to leave out a snack for Santa. He gets very hungry delivering all those presents – what would he like, do you think?'

'Yogurt.'

Gurl bursts out laughing.

'Mummy? Is that the Gurl laughing?'

'Yes, that's Gurl.'

'Is she three?'

'Jealous of you,' Scarlett whispers to Gurl. 'No, Fintan, she is a very big girl. In fact, Gurl is a woman. She's not really a girl anymore. You want to see how grown up she is? You mind, Gurl? Press there.'

Gurl leans in, waves. 'Hi there, Fintan. This is Gurl.'

'I want to see my mummy.'

'Sure you do,' Gurl twists the screen to Scarlett who waves.

'Mummy, is Gurl coming with you?'

'No, it would be too far.'

Frank chuckles. 'OK, Fintan, let's go find your furry stocking.'

'And I'll call you later for your bedtime story, Fintan.'

'Come home now. Love you. Kiss kiss.'

Silence falls. The screen dulls. Gurl tucks in her chin and holds. Twists her head from side to head and rolls her shoulders.

'He's real cute.'

Scarlett nods, blinking now against the hollow collapse in her chest, the longing to scoop him up. And again, that feeling that the car is static and time is unspooling at high speed, that she has mislaid the heart of things. There's a white flicker to the left of the windscreen and she cranes over the wheel and yes, there is a white predator bird overhead, tracking in purposeful navigation. She decelerates, breathing through her teeth against the battering in her chest, glaring at the straight road, now terrifying in its uniformity. She glances again at the bird powering ahead: is it guide or siren, leader or follower? And what is this icy trickle of fear crawling up her spine?

'Steady, now,' says Gurl, as if she has picked up on this signal of unease. 'You'll be home real soon.'

Scarlett smiles. 'Yes, home soon.' Warmth returns, the grey fades and she picks up speed. Recovery and restoration, she murmurs to herself: move forward with intent.

'Tell me what it was like when you knew Fintan was coming,' Gurl says softly, turning sideways.

'My ovarian reserve was 5.5 per cent.'

'Your what?'

'How many viable eggs I had left.'

'You can measure that?'

'Estimate, with a pretty fine margin of error. So the chances of that moment were 3.4 per cent. Which is *exactly* what I was thinking the first time Fintan smiled. I can see him still. Twelve weeks old, in his little bouncer on the floor in front of me. I am lying on the couch and pressing the bouncer with my foot – he likes this, the gentle rocking up and down, and he falls asleep that way, slips into that cosy sleep. He's blinking, his little pink mouth yawns. He's staring with the big blue eyes and I'm smiling back at him, I put out my finger so he can wrap his hand round mine, it's a hot grip and I just cannot believe the mystery of it all. Twelve weeks ago, he was what was inside me – and even then I could not quite believe in this possible world. I think of what it takes to get right to this moment – all the genetic permutations, the moving parts, the chromosomal event, the casual plot twists that at any second could have stopped him in his tracks. Me, Frank, our parents, that moment of sex that happened just when conditions were perfect – the statistical odds of this moment.'

Scarlett shakes her head, remembering. 'Then his lips part pinkly, the blue irises twinkle – I swear it's not the light, it's a December dark, clouds huddled thick and low at the window, cars hissing by in the rain outside – and this boy, my boy, my son is smiling, and I'm laughing out loud, and he giggles and smiles all the more, and wiggles his hands, little fists bouncing in the air at his own discovery.'

There was Fintan and the whole world came down to him. She feels a regret now for all the moments since where she has failed to linger. Forty-one months already – 1,214 days, 1,748,160 seconds – and so many of them forgotten! But she smiles now at the glassed road and a state she could not have imagined. A delirium that sets up. Poleaxed by a bewildered, astonishing love.

The night he was born, hours after Frank had gone home, she lay alone on the sixth floor in the hospital bed, with Fintan swaddled by her side. Swaddled blue, the traffic hum at 3 a.m. They had told her she would not sleep after the epidural and she was indeed super-alert, mesmerised by his tiny lipbuds. Looked up and imagined she saw her former, pre-birth, self, leaning against the doorframe, smiling and waving goodbye. To one-ness. She slid forward into what still seems like the most natural of futures. Each day, Fintan sheds a skin and slithers into the next development stage. Faster, stronger, smarter. His little feet that once were thumb-sized pedal furiously along the path. In less than twenty-four hours she will see him and find him, again transformed. It is always that way. This beloved boy and forever companion that they will have and hold and then let go into a world magnificently enriched. It is the grandest enchantment.

Scarlett glances smiling at Gurl who is frowning at the forest blur, knuckle-kneading her thigh.

'Oh, I'm sorry,' Scarlett's hand flies out, too late to snatch back her own insensitivity.

'Hey, you love your little boy, that's cool.' Gurl swats away the apology and leans forward to rummage in her tote.

'You want some iKandy?' She rattles a glittery pill box.

'No thanks. I don't like mood regulators.'

'The fix is real good,' Gurl pops one in her mouth. 'No side effects. All legal. You never had one?'

'No.'

Gurl chews, considering. Scarlett thinks of the present they'd hidden in the laundry closet the day before she flew. A big galley ship with canons and a dragon mast and sails made from real sailcloth, a treasure chest and a cage for prisoners. Each time they go to the toy store, Fintan stands staring at the box and asks her to read out the names that he now knows by heart.

'Tork, that's my best one,' he taps the box, pointed at the charcoal warrior with a silver helmet, a tuft of red beard, his hand a long spear. A shield curved over his heart.

'You could be Fintanvagen,' she says. 'Fintan the Fintanvagen. You can be the leader.' He traces his finger over Tork, turning over the responsibility of leadership.

In the playground, she sat with Frank on the bench and watched these life rehearsals unfold in the sandpit. When some kid pushed ahead, Fintan holds still and observant.

'He's a watcher,' she leant into Frank's shoulder. 'He does not rush in. Think first, act later. Much like his Dad.'

'First time with you, I acted,' he kissed her hair. The October sun dazzled her eyes.

'Eventually.'

'Only took a couple of hours.'

'I was easy.'

'That hotel – '

'With the view of the palace – '

'Of the wall of the palace.'

'Twenty-second floor.'

'But the room was 2108.'

Is this how it would always be, she wondered, sitting on the bench. Will they be filling in each other's sentences, dovetailing shared remembrance, building this jigsaw of affection. Celebrate your anniversaries. Make a ritual of those things to remind yourselves of what you have, the gift you are to each other. Would they have to work at it? She could smell him, knows the contour of Frank's knee, how the skin on his knuckle creases. How do you measure intimacy? What is its intrinsic value? When she tries to imagine his absence, she can glimpse the size in the shape of the day, the passing hour.

'You were still grappling with chopsticks,' she said.

'And you were looking for a reason to leave,' Frank squeezed her arm. She closes her eyes and cries a little for all that can be lost in an instant, the fragility of it all. Fintan there running in the sandpit. Beside him a girl in a lilac top addresses figurines balanced on an upended bucket. To the right, a toddler staggered forwards clutching a blue plastic spade. Everywhere the melody of new family, early years.

'You dropped off,' she prodded his arm.

Ah, the luxury of sleep. Which is what they do these days when they are not with Fintan or playing catch-up with the life they have assembled. This reconfiguration of family, the new centre point, now the beating heart. Now, adrift after three years of mummy and daddy they have arrived exhausted and re-formed. Unpeeled. Parenting strips away the varnish, rubs you down to the bare wood. And what is lost is what you once were. An irreversible state of vulnerability. There is a new variable in every equation.

She swept her hand across the slides, the wooden huts the admiring parents looking on astonished at the brightly coloured offspring finding their place in the city.

'There's no way back, you know.'

'To what?' Frank murmured.

'To being non-parents. I mean, even if we lost him.'

'You are so fucking literal sometimes.' Frank tipped her head roughly from his shoulder and stood up.

'I only meant – '

'Do you have to articulate your morbid thoughts? Do you have to ruin all this?' He flung his hand across the playground, the bobble hats and balloons, buckets and spades, the swinging children, the parked buggies, the cityscape of young families on a Sunday afternoon.

'You don't know how to share, do you?' Gurl's voice snaps low. 'I know you're sitting there at the wheel, remembering stuff, stories about your life that you never tell. Still you suck up all mine.'

'What's your problem now?' Scarlett turns, but Gurl does not meet her gaze – just stares straight ahead.

'Does it make you happy, having a nice car?' Gurl runs her finger slowly along the door trim.

'Not this one,' Scarlett smiles. 'Why d'you ask?'

'You know, just cos we've both got pussies don't mean you and me got anything in common. Schooling and money gives you the edge. Makes you more of an important human than me. One of those making the future.'

'You're saying this, Gurl, not me.'

'Ever cleaned out your own greasy kitchen? Or someone else's shower tray? Cleaned shit, puke – '

'I'm a mum, remember.'

'You're the one with the big job.'

'So hairdressing was your only option?'

'Oh sure, I guess I coulda done more than chew gum and doodle on my class books and listen to some teacher who didn't give a fuck. I coulda not been a wannabe porn star and not got expelled and not got pregnant. I coulda arranged it so I was born like you. Into a family that was conducive to bettering myself.' She runs her fingers over the console like she is testing for imperfections. 'And now you're thinking cos I know that word that proves I coulda done different.'

'You're saying all this, Gurl, not me.'

'Thought you liked doing things for yourself?'

'I work. I look after Fintan when I'm at home.'

'You cook his food?'

'The nanny cooks.'

'The human nanny Frankdaddy doesn't want.'

'She prepares it and I eat with him in the evenings.'

'You clean your own bathroom?'

'The answer to all your questions is I either pay people or buy machines to do domestic tasks, so I can spend my time with Fintan.' Scarlett's cheeks heat up above her stinging jaw. She lowers the air con. 'Why don't you just spit it out, Gurl, and tell me what's got you all riled now.'

'Why is your hour is worth more than mine?'

'Read Karl Marx.'

'You're growling.' Scarlett hears the smile in Gurl's voice. A wheedling tease, designed to push and needle.

'I create something valuable for a business that I helped to build. So, I get a share of that.'

'Bet that share is a real big number. Bet it's millions.' Gurl slaps the dash.

'You want me to apologise for having a good job, a great education? You want me to apologise for your life?'

'I just wanna know what it feels like to have so much.'

'No one ever tells the truth about money.' Scarlett shrugs.

'What you all het up for now? All your guilt, not mine.' Gurl pushes out her lower lip.

'And I'm starving. Where's that diner?'

NIGHT

Gurl barges through the door with a high-heeled urgency and maximum noise; swinging her tote, striking short sharp strides across the floorboards. The barman looks up, clocks her heels, carries on polishing the glass. Two guys a stool's width apart swivel their heads. Gurl stops dead, glares at the stool hounds who turn back to the violent natural drama unfolding overhead: a close-up slow-mo of bloodied teeth ripping out the throat of an unidentifiable prey. On the far side of the bar, a fat woman sucks on a beer. Behind her, a pool table waits in the shadows.

Gurl dumps her tote on the floor like a toolbox and smacks the stool beside her. Scarlett sits.

'Afternoon, ladies,' the barman spreads two lean arms on the counter.

'My favourite bit,' says Gurl.

'What's that?' asks Scarlett.

'*Ladies*,' she repeats. The barman smiles. Scarlett takes in the dark hair, green eyes.

Gurl studies her effect in the mirror behind the bar. Cocks her head to one side and narrows her eyes. Then flicks, lizard fast, to the stool guy watching, who snaps back to the wildlife. The second man studies his spread hands like he's checking for tremor.

'Told you it was retro,' says Gurl. Scarlett nods at the silvered sweep of mirror, the bottles gleaming silver and green and blue, the scent of bleach and beer and a salty undertone that calls to mind the sting of margarita. On a small screen behind the bar, a massive long-horned bull lies chewing nonchalantly on the sidelines, a red and white scarf round his neck.

The barman slides two mats in front of them.

'You Tom?' Gurl points at the gold banner on the mirror: LLIRG & RAB S'MOT.

'No ma'am.'

'Then you can be Tom today.'

A smile tugs at his lips. 'What can I get you ladies?'

'Beer for me. And for my friend here?' she jerks her head at Scarlett.

'I don't know.'

'Now, Tom,' Gurl leans confidentially into the counter. 'Scarlett here is taking lessons in how to live a little.'

Tom folds his arms and nods gravely. He reaches down to the fridge, snaps the bottle cap, places a lime slice in the neck, and places it on Gurl's mat.

'And for you, Ma'am?'

Scarlett frowns at the array of bottles. 'I'm thinking – '

'You surely are.'

'I'm driving.'

'So take a Chaperone like everyone else. Tom, you got the pills, doncha?'

'Sure do.'

'I'm not in the habit of taking that stuff.'

'Well you need to start, Scarlett, cos that pill is only about the best invention of modern times. Folks can go out have a few beers, take a Chaperone, sober up, no fighting, no swearing, no crashing cars. Everyone having a good time being sociable. What do you say, Tom? You look like a reasonable man to me. How much bar time you served?'

'Been here since I was eighteen.'

'Lemme guess how many years that is. C'mere.' Tom leans across the counter and she peers deep into his green eyes. The stool hounds stare. Tom holds her steady gaze.

'Ten years,' she pulls back. 'You're twenty-eight.'

'Gimme five,' says Tom. 'Yeeeah,' the men chuckle. The fat woman nods approvingly.

'And you got good clear irises, too.' Gurl slugs on her beer.

'So tell me, Tom, in your professional opinion and your ten years barside, would you say that Chaperone has made a positive contribution to sociable drinking?'

'Yes, ma'am.'

'There's your expert evidence, Scarlett.'

'I'm not sure what I want.'

'See there's your problem, Scarlett! At any moment, at any time of day or night, I could tell you what I wanted. And here you are overthinking every little step. So, Tom, could you recommend a suitable drink for a woman who can't decide?'

'You like a short or a long, Ma'am?'

'Long.'

He studies her – as if he is taking a reading from her face. It is the same clear-eyed gaze as Gurl's, as if they were kindred spirits, and for a moment Scarlett has the impression she is straddling parallel worlds.

'Can I recommend an Ocean Breeze. Shot of Absolut with cranberry grapefruit.'

'I'll have some water.'

'She'll have the goddam Breeze,' Gurl waves a hand. 'Just behave, Scarlett. This here is a social occasion.'

The barman busies the ice. 'So tell me, Tom – '

'He ain't really Tom,' says the bar hound. Gurl leans back on her stool and offers him a challenging look.

'Tell her you ain't Tom, Tom?' he winks.

'I am whatever the lady wants me to be,' Tom smiles.

'Al-right,' Gurl applauds his gallantry. 'Now I am hungrier than I have been in a long time. How long to get something to eat round here?'

'Kitchen's not so busy right now. Won't take long. Got some real good crab cakes on the go.'

'Yeah, real good,' says the fat woman loudly from the other side. And her plump hand drifts up to her mouth, as if she too is startled by the sound of her own voice.

'I'll have the crab cake,' Scarlett smiles.

'Make it two, Tom, and make it snappy. Old Gurl is fit to starve to death right here on this stool.'

'What age do you think I am, Gurl?' Scarlett says to her reflection in the bar mirror.

'Whoa, I ain't falling for that one.'

'Tell me, I'm interested in the truth. How old do I really look?'

Gurl reaches out and turns Scarlett's chin. 'Well.' She studies her face with a professional frown. 'You ain't wearing much and I can't see you've had work done. Don't seem like retro girl kinda thing.'

'Correct.'

'You look bit on the starved side. Worked out, antsy. Like you're always running from somewhere to the other.'

Scarlett grimaces. 'Thanks.'

'Hey, you asked. You coulda just told me your age instead of fishing. Anyways,' she turns away and picks up her bottle. 'I'm gonna say you've cruised through the big four-oh.'

'I'm thirty-nine.' Scarlett sighs and picks up her Breeze.

'I could give you a whole makeover, for sure,' Gurl squeezes the lime. 'Soften you up. Start with that Frankdaddy and kick him into touch. Where he gets off making you feel guilty back there.'

Scarlett laughs. 'He's a good man, really. He's just worried I won't get back for Christmas.'

'A good man who makes you feel bad. Now how do I know *that* feeling.' Gurl slaps her knee. 'OK, your turn. You tell me something true about how *I* look. Something someone else won't tell me.'

'Like something even Roxanne wouldn't say?'

'Specially Roxanne. She is kindness itself, never says a hurting word.' Scarlett looks her up and down struggles for an observation that can be packaged into words that will not bruise. There's a tatteredness that is appealing now – a little tear in the laced dress frill and there's the bald boot toes – but Gurl is otherwise revitalised by her three-minute make-up in the car park, the soft bar-light sparkle, the iKandy, the beer, and the prospect of food.

'OK, how about this: your puffer is vile.'

Gurl gasps, looks down at the jacket hanging on its hook. 'That poor puffer that keeps me warm and cosy? Pretty in pink, Blane says.'

'Well, he's wrong. It's the colour of raw meat.'

'What colour should it be?'

'Ice blue,' Scarlett taps the bar. 'Show off your eyes. Which are lovely, by the way.'

'Sure are,' Gurl flutters her lashes in the mirror. 'My best feature.'

A white-shirted waitress sails through the swing doors with a large circular tray balanced on her shoulder.

'Didn't realise how hungry I was,' Scarlett murmurs, for there is something uplifting about the arrival of the white square platters. Golden squares of crab cake. A giant green gherkin speared with what looks like a hunting knife, as it if had been making a bid for freedom.

'Tell me about my makeover,' Scarlett picks up her fork.

'Start with this,' Gurl flicks the black strand. 'How much you pay for that cut?'

'Too much.'

'They seen you coming. Sure, it's sharp and fine, but who-ever is styling you – '

'Mika.'

'Ever had it short, or you always keep it up like that?' Gurl raises a forkful of crab. 'Cos when you're pinning it up, what you need is a restyle.'

'Just can't decide.'

'That Mika ain't even looking at you anymore, Scarlett. And neither are you. You got the off-the-shelf working mom look. The don't care hair, I call it. That bob is between here and nothing. And that Mika will keep you like that for ten years. Head like yours you got two choices. One: you go right on the jawline, razor sharp. Stick in a fringe it will cut you like a knife.' Gurl touches the midpoint of Scarlett's neck. 'You got the cheekbones,' she frowns, 'hmm – yeah maybe – but you ain't got the time. Now that's a high-maintenance look. Needs an hour on make-up, so it's for the pros.'

She snips at the gherkin, studying Scarlett all the while. 'Option two, and this is the right one. Take it up an inch, just skimming the shoulder and you get a little kink outwards. *Not* in. Make you look a bit more goddam interesting. And, boy, do you need that!'

Scarlett experiments with twirling a strand of hair outwards.

'The kick says: I can lighten up. I can party.'

Scarlett laughs. 'Oh wow, you've got the wrong idea – you should have seen me a few years back.'

'Hair has gotta bring out the private you. The you that you dream of being. If I had my scissors I'd do it for you right this minute.'

'Maybe one day I'll come to your salon.'

'It's a new place, now, more upscale – though I'm part-time on account of looking after Roxanne. Clients got money and they *love* Gurl's stories.'

'Why'd you pick hairdressing in the first place?' Scarlett pushes her plate to one side and Tom appears immediately to take it away.

'Picked me. Remember I got kicked outta school when I was sixteen. My mom worked down the salon and I'd been hanging round since I was a kid. Not a real salon, more like a shed with a water supply. A shed full of loser women sitting around telling stories about their loser men. Place where they could go to get away from kids and men so they can dream.' Gurl covers her plate with the napkin. 'They loved me cos I made them laugh with my stories.' Gurl tilts her beer back with a smile. 'Couldn't get enough of the stories, my salon ladies. Hairdressing is skill and psychology,' she taps her temple. 'I can tell straight off looking at a woman's hair what she needs. Ain't your real self you see in the mirror, you know. You see what you feel. You start imagining a different life, then you might actually find it!'

Gurl swivels round to contemplate the empty floorboards behind them that stretch out to the far wall. 'Got themselves a cute little dance floor here.'

'I always wanted to dance,' Scarlett twirls the Breeze, a watery melt of pink, like frozen plasma. 'But it's hard to be elegant when you're tall.'

'Bullshit,' Gurl snorts. 'Dancing means letting go. Like diving. Like falling in love. Just be the body.' Gurl twists back, clunks her bottle to Scarlett's glass. 'And I'm gonna show you how. She hops off her stool and dusts down the frill of her dress and takes Scarlett's hand. Who pauses, glances over at the bar hounds.

'Hell with them. This is your once in a lifetime opportunity,' Gurl grips her wrist.

'I can't.'

'Lemme give you this one thing, Scarlett. My way of thanking you for being a one-day friend.'

Scarlett cannot refuse the offer and the grip. She steps down and Gurl spins round. 'Here Tom,' she holds out her phone, 'Stick on Gurl's list five there.'

Gurl takes up position behind Scarlett who looks down at the floorboards. She feels Gurl lift both her arms from behind and let them flop.

'Re-lax! Close your eyes,' Scarlett obliges. 'Loosen your hands.' Gurl flaps Scarlett's arms back and forth with surprising force. 'Let 'em go' – she pulls both shoulders back.

'Tom, we need some volume here,' she calls out. 'Scarlett, we're gonna sway with the song,' she places both hands lightly on Scarlett's waist.

'I'm feeling lucky O – O – O. I'm feeling luck-eeeeeee,' Gurl sings, strong and clear.

'Hey, you,' she yells at the bar hound gawping on his stool. 'Ain't you got some killing to watch?' She points to the screen. 'This here's a private dance class. Show's later.' And he twists obediently back to the bar.

'Close your eyes, Scarlett: you are just the music. And you gotta sing along, too. Swaying and singing.'

'Oh OH OH I'm feelin luck-eeeeee,' Scarlett tries, private and low.

'You got to be the music. 1 – 2 – 3 – 4 and left, look straight ahead. Line dancing gonna work for you, I know it,' Gurl stops and dusts her hands. 'Give you some numbers to think about. Watch this,' she takes Scarlett's hand. '1 – 2 – 3 heel,' she taps her left bald-toed boot against the right heel. 'See that?' And Scarlett does, at least she feels the tug like a fishing line pulling her along. '1 – 2 – 3 heel,' she mouths to the comforting rhythm, numbers cleaner than notes.

'I'm feelin luck-eeee Oh Oh Oh,' Gurl sings. 'Good, that's real good, you're dancing Scarlett, now keep the line and we gonna turn.'

Right to left they go again and again Scarlett feels a faint surrender to the physical, like she has stumbled on a new sense, but when she turns her head she catches the waitress's sceptical look, arms folded over her crisp shirt. Scarlett falters. The moment is lost.

'You did good, Scarlett,' Gurl reels her in. 'Gonna be dancing that line 'fore we leave this place.' The music fades out. 'Now you get to work on that Breeze. I got something you wanna see.'

'Number seven, Tom,' she calls, 'Nice and loud now – show's starting, people!'

Girls flings off her shoes and begins to unbutton her dress.

'Yee hah!' whoops a stool hound. Gurl whips her head round and fixes him with a stern glare. She slips off the dress to reveal a black leotard and arranges herself in opening pose: arms raised, arched back, each rib bone a clean curve. Left leg extended out behind her, bare foot arched so hard it cramps to watch. Her whole weight, all of Gurl, exquisitely balanced on her bent right leg, foot planted outwards. There are words for this precision engineering, but it is not a language Scarlett knows. The lights begin to fade, the bar is cast in cinematic darkness but for the flickering screen glow and Gurl's spotlit dance floor.

She dips, her hair brushing the boards. The song spirals and Gurl unfurls, lets the music pull her upwards, her torso rising like a marionette powered by melody. Raises her head and twirls so fast her feet are a single pivot. She slows and then accelerates into a black spinning pole, her hair a blur of angel dust. She spins and soars an impossible parabola and lands, and the song unwinds and her feet grace the boards, a dust cloud shimmers. She jumps again, and for a second seems to hover in space before she sinks into the final note. A fold of limbs, a streak of black and pale gold, legs split on the boards, arms wide, face wet and fresh and radiant.

All rise cheering. Gurl stands up and bows. Tom applauds behind the bar. 'Sure is the classiest moment to ever grace this place.'

'And now,' Gurl beckons Scarlett, 'Come on y'all and line up. Tom, you know what we gotta have? The Margarita Girl.' She pulls Scarlett into the middle of the floor.

'Keep it simple. Just fall into the numbers. You can do this.'

'Are we ready to take the line?' yells Scarlett above the swell, hands raised above her head. The fat woman slithers off her stool and trundles over. Tilts her chin from its rolls, majestic and defiant. And the three women slip solemnly into step in perfectly matched time.

Skintight jeans
Lookin' kinda mean
Crazy Margarita Girl ...

The song pulls them right and left and back and forth and they're claiming the space, keeping the line, Gurl at the centre, Scarlett on the left and the fat woman on the right, flowing and advancing and falling in graceful union. Scarlett's black hair loose about her shoulders, Gurl's luminous wisps and the fat woman's chins quivering to the beat.

'You can dance, Scarlett. You're a dancer now!'

Scarlett's keeps her eyes locked on the mirror lettering, she is swept up in rhythm, her ears at one with the music, and there's something like a burning torch soaring inside her chest wall like it could burst right out of her mouth.

'Take a bow, ladies,' yells Tom, and they do. The bar hounds clap and whoop. Scarlett slaps the notes on the counter and picks up her parka. Gurl slips on her dress, grabs her tote and the pink puffer, and they take their leave, laughing and giggling, arm in arm, flinging air kisses to the standing ovation as they back out through the swing doors.

Outside, the raw air hits them. The Buick glimmers frosted gold beneath the lights. Scarlett pulls on her parka, feels the sweat on her back cool to ice. Gurl shrugs quickly into the pink puffer, her gloves are blood red in the floodlights.

'Hallelujah, I can dance!' Scarlett goes, and Gurl grabs her lapels and yanks her down so that their foreheads almost butt. Her lips are hot and hard, Scarlett tastes beer and saltiness in this small mouth, her tongue is a quick flicker, and then she is pushed back.

'And you can kiss a girl, too,' says Gurl, eyes sparkling and triumphant. Scarlett watches her lead the way across the car park like a catwalk pony, short steps on solid ground.

'Just so you know,' Scarlett calls out and Gurl spins round. 'I still fuck Frankdaddy.'

And Gurl laughs – a big raucous crack in the black night.

Scarlett docks her phone. 'I have to call Fintan.'

'Mummy!' he greets her. 'Daddy put a beer for Santa! And my yogurt.'

'Mmm hmm, beer and yogurt, Santa will gobble that up!' Frank chuckles in the background.

'Mummy, where's Gurl?'

Scarlett nods to Gurl, 'Hey Fintan! This is Gurl and I'm hearing Santa is on the move.'

'My Daddy has Santa on the screen. He is – Daddy, where is he now?'

'He'll be leaving the North Pole very soon. See? That flashing beacon right there? That's the sleigh getting packed up. Mummy is much further north, so if she looks up at the sky right now, she might be able to see him.'

'I'm looking.' Scarlett leans, unrolls the window and twists her head upwards. Stars sparkle behind the thinning cloud, as if the sky has begun to shyly disrobe under cover of night. She wants desperately to find what could be presented as evidence, scans quickly in the direction of the airport and there, freshly revealed by a drifting cloud, is a pulsing white light. 'YES! I think I see Santa's sleigh!'

'Hawww,' Fintan gasps, she imagines his wide-eyed wonder. The light pulls quickly away eastwards, a planeload of just-in-time homeward bounds like her.

'Gurl?' Fintan's voice loud and breathy fills the car. 'What ummm – what did you ask Santa?'

'I asked him to get Fintan the biggest best Christmas gift ever! You know what I said? I said, "Father Christmas, you gotta bring Fintan's mom back home with a big red bow on her head!"'

'Can't wait to see that bow!' says Frank. Fintan giggles. Scarlett smiles at the sound of togetherness and snuggles. And a memory trembles like a mirage on the windscreen – Christmas morning, Frank naked in bed beneath wrapping paper. She cannot place it specifically in the timeline. Somewhere in the Old Past of Two that has been obscured by the New Past of Parenting. Was it all too soon? Was three years of Two enough? A throb of regret, a nostalgia for the unrecoverable past, and the pain of longing has the quality of dull earache. There is urgent repair work to be done with her and Frank and all that has been lost.

'I'm gonna leave you to talk,' Gurl opens the car door.

'It's freezing out there!'

'Got my favourite puffer!' she grins. 'I gotta call Blane.'

'Mummy, story!'

Scarlett leans back into the head rest. 'You all snuggled up?' She listens to his breathing, the blanket rustle, and a faint scratching sound: Fintan running his fingernail along the cushion cover. She pictures the bike-shaped present in the basement, the red and silver wrapping with a white sash. 'One day,' she begins slowly, 'Fintan is riding along on his bike and – '

'A big bike!'

'With shiny spokes that sparkle in the sun and two big wheels and some teeny stabiliser wheels at the back. So whenever Fintan wobbles, the little wheels stop his fall. But guess what? Very soon he will be able to catch his own wobble!'

'Wobble, bobble.'

'Then one day Mummy and Fintan are in the park with Buster. Who is running around sniffing dog's bottoms.'

A cascade of squeals, Frank's 'eeeyuck' and Fintan's 'eeeeeeeeyuck.'

'And Fintan is riding along the path on the big shiny bike. And the sun is shining, and the clouds are puffy, and the grass is green, and the dogs are barking – '

'Ruff ruff.'

'And you are wearing a blue jumper that you haven't got yet because you're bigger in this story. So you're rolling along on your big bike with the shiny spokes and the little wheels. And then you have a wobble! But ... guess what? Fintan doesn't fall. Then Mummy says, "I think it's time to take off those wheels." And Fintan says – what do you say?'

'YESSSS.'

'So Mummy takes a screwdriver out of her pocket and unscrews the little wheels and takes them off. And she says, "Fintan are you ready for a big adventure?" '

'YESSSS!'

'And Mummy gives you the biggest hugest hug ever. Bigger than the biggest bear hug! Big as an elephant, wide as an albatross. And you sit on the saddle and give Mummy a huge wave, give the horn a little toot – '

'Toot toot.'

'And Mummy watches you roll forwards. You take a teeny wobble but you catch it and then you speed up till you're fast as a bird. And Buster has to run to keep up with you. And everyone in the whole park is clapping and shouting yippee! And on you go, Fintan, riding right into the great big adventure. And Mummy is right behind, cheering you on.'

'Now big boy,' says Frank, 'time for bed.'

She hears the sound of Frank's tread to the hallway, with Fintan singing. She remembers unhooking the guide pole from his tricycle. His helmeted, red-cheeked insistence. *Mummy no push.* What he wanted most of all is to go solo. Frank runs ahead down the slope, Fintan sails off across the flat, smiling grandly. No brakes, he can't stop but either one of them can easily skip ahead and do it for him. The little blue trike comes to a clattering halt and she kneels down to gather him in, feels his heart beat puppyfast. All these first steps point towards his own horizon. The whoop of acceleration, the thrill of letting go. Freewheeling towards his future.

Did the venture, Mummy. Yes! You did it, best boy! Frank ruffles his hair and says *ready to go again*? Fintan climbs back on, and passers-by smile at the little boy in the blue jumper. He is blessed, all three of them are blessed there on the path in the park, held in that forever memory.

'Frank,' she smiles into the phone, 'I was just remembering when we took the pole off his trike. Seems like only yesterday.'

'Another big milestone coming up,' he laughs.

'Listen, I'm sorry about – '

'Hey, sshh, everything's fine. I've been down to the basement to check on the you-know-what.'

'Oh!' her heart sings.

'Had to sort out that white sash thing. You better keep up the day job because you'd never cut it as Santa's elf.'

Her smile spreads wide in the darkness. 'Frank, he's going to love that bike.'

'He's going to love you coming home even more.'

'Double-whammy Christmas.'

'For all of us,' he says. And true to form, Frank does not say more. They are both held close and contained now in this fibre-optic moment, bound by cables submerged five thousand metres beneath the ocean. Scarlett imagines a pale pink squid in elegant sidestroke, drifting over their worldless communion. A speckled sand crab scurrying across the seabed where the cable sleeps in its hermetic steel tube, only a metre below the scuttling claws. And she smiles warmly, at the vast undersea network that keeps them together, unites them in the sub-marine transmission of this complex exchange. Marvels, yet again, at the splendour of human ingenuity, the technological enrichment that amplifies and underpins the human experience. These shared mute seconds are the conversational heart, the meaning that beats in the gaps, in all that goes unsaid and uncoded between two people.

'You all set?' Frank punctures the pause.

'I'll be at the airport soon, call you when I board.'

Silence rings loud now in the Buick's chamber. In the rear view, Gurl talks into the phone, pacing briskly back and forth between the fire hydrant and the dumpster. Every now and then she stops in time with the conversation, the gesture of argument and frustration. A figure darts out from the rear of the bar, pauses, glancing at Gurl, then hurries away, climbs into a truck and pulls away.

Scarlett considers the clock: her Volo deadline is ten minutes ago. Colin will be fuming at her churlish delay.

'Hi there.'

Scarlett starts. Stares at her speaking phone. 'Hey?' says the familiar voice. 'Are you there?'

'How the fuck did you do that!' Scarlett grabs the phone from the dock.

'I hacked you,' says FlyBoy.

'Yes, I can see that but – what – is something wrong?'

'Nope. No emergency.' His voice light and airy.

'So why did you do this?'

'I thought you'd appreciate it.'

'Why would I appreciate anyone hacking my phone?'

'I wanted to tell you I found your Easter egg.'

'Aaah,' Scarlett flops back relieved. 'Clever boy.'

'I was doing a fix and I dug up an egg at Line 313. So I opened it – which didn't take too long – and I saw what you did six months ago. You wrote yourself into Volo's code.' She grins as FlyBoy describes his discovery. 'So, every night at 22:47 Volo sends you a secure mail. A routine command that no one notices. And the message is, "Regression sequence concluded." Undetectable by anyone.'

'Apart from you.'

'So why? Why d'you do it?'

'When you were a kid, did you ever put your handprint in wet cement on the road? Or trace your initials? The signature of the creator carved into the structure. It's not that strange to want to leave a mark, so you can be remembered forever.'

'Never took you for that kinda geek,' says FlyBoy.

'It's very human, you know, to fall in love with your work. The masons in ancient Egypt used to etch the stones, so they could live on in the architecture.'

'So ...' FlyBoy pauses. She pictures his pale streak against the lime green. 'You wrote yourself into the code, and you hid the egg because you knew that one day you'd be waving goodbye to Volo. And you, like, want to stay in touch?'

Scarlett looks out at Gurl; the puffer neon pink in the headlights. 'Volo's email is like getting a card from an old lover. It's a little reminder of a shared history. But eventually we have to say goodbye to everything we love.'

'That is so weird!'

'We all have our own ways of being individual.'

A tinkling sound as FlyBoy taps the keyboard.

'Well,' she says, 'I guess you are just hours away from finding my other Easter egg?'

'Already there. Line 222 in Block 27. That's why I'm calling. I can't open this one. Can't get past the encryption.'

'I take that as a compliment.'

'I mean I *will* get past it. But it'll take a bit of time. And since we have a new owner – '

'Not yet,' Scarlett raps the wheel.

'It's a done deal.'

'How can you be so sure?'

'Cos you will say yes. I know you want to see Volo go all the way.'

'I am worried about what the buyer might do. Could do.'

'But it's happening anyway,' says FlyBoy.

He is impatient to move on. She knows that ethical concerns do not engage, he views them as a kind of indulgence, a luxury, and an obfuscation. He does not consider himself responsible or complicit. He does not worry about unintended consequences. He is focused on possibility, design, implementation. The new handlers will give him endless rope and he will take it all the way. FlyBoy is in pursuit of a single goal and it will be up to others to decide on application. While she, Scarlett, is distracted, transfixed by the tension between pros and cons. And risks sinking into paralysis.

'So, like I said, I could crack this egg but since it's yours, I thought, well – '

'You didn't want to crack me open without calling me first?' Scarlett laughs. 'Oh, I am touched.'

'I need to see what's inside.'

'So you can axe it?'

'I want clean code,' FlyBoy sighs. 'Easter eggs are not convention anymore, you know that. You'll be gone and I'll still be here. So you need to let me in.'

'There's a poem inside the egg.'

'What kind of poem?'

'You mean is it an incendiary device that's going to unleash a worm? No!'

'Then what's it do?'

'It doesn't *do* anything. It's for thinking and feeling. It's a touchstone for humanity – have you ever read a poem?'

'At school.' FlyBoy says tersely. 'Nothing happened.'

'Hang on,' she taps a series of verifications and prints again. Presses SEND. 'Here, go read it. Now I have to call Felix.'

She kills the line. All her work conversations reach a natural terminus and never arrive at the formality of goodbye – except, of course, with Felix, who preserves the structure of closure.

Gurl moves to the front of the Buick. She stands tapping her leg impatiently, head crooked into the phone, her free hand tucked inside the puffer. She catches Scarlett's gaze and rolls her eyes, holds up a splayed hand: five minutes.

Scarlett nods. Of course, FlyBoy does not yet know that she has hidden something else in Lab. In Vault, the place where all the coders refuse to go since they believe it's haunted by the ghost of their predecessor, the celebrity coder who chose the empty server cage of wire mesh as his death chamber and fitted a removable bolt that fastened from the inside, so that he could not be rescued.

She has watched the four-and-a-half hours of footage recorded by the internal cameras that show his horrified colleagues battering uselessly at Vault's security door: high-grade steel, 160 mm thick and fitted with three anti-explosive devices. He had reprogrammed the electronics and they could not crack the code in time. Instead, they stepped back and joined the millions who watched his suicide broadcast, howling and swearing at the big screen as he bled out five metres away in his customised shrine. Arguing and yelling about whether to respect his wishes or call the emergency services – which they eventually did – and then calmed, exhausted, they sobbed their way to the closing scene: a choreographed crumpled still life, a crimson pool on the pristine concrete floor.

She has not told FlyBoy that she has left her mark in Vault: in the wall safe that everyone has forgotten at the very back, on the master document hard copy where Volo's code rests like a sacred text. She has annotated the 21,802 pages with little comments in the margin, transformed it into a living document that tells the story of a human project, for their journey from dream to execution is the story of the birth of an intellectual property. The human history of creation. One day FlyBoy will find it and consider it yet another of her eccentricities: the romanticisation of process, her emotional attachment to work. He will turn her pages and shake his head and she will think of him as she will always remember him, palely cast against the lime walls of test, his bony white arm holding Volo's gleaming sphere up to the light.

She pictures the server lights blinking reassuringly in Vault and Rand returns again, for Scarlett too has always felt that 'joyous sense of confidence when looking at machines … every part of the motors was an embodied answer to "Why?" and "What for?"' Oh, but now she needs to get better at graceful exits! The truth is, she has not been entirely honest about her objections. Seduced by the wonder of invention, she has allowed affection for Volo to cloud her judgement. It is time for the work to be entrusted to those who can realise its potential. *You and Felix will still have your fireside chats about ethics and engineering*, said Frankisalwaysright. Time to say goodbye to all that and accept the mourning period, the grief of letting go.

And Project Fintan is a work in progress. This is the kind of thought that doesn't travel well in articulation but has the bloodlessness of truth. Parenting is a process of leave-taking and one day she will have to let Fintan go: in the meantime, he must be equipped with the tools to walk into the future and find his own way. She believes he will emerge unscathed by nurture, since theirs is a healthy balance – the see-saw of Frankisalwaysright and Scarlett, the container of uncertainty.

Fintan will need to figure out which advice to discard for so much wisdom decomposes like blackened mulch that should be ground underfoot. A student of the past can become a blinkered horse, yet nostalgia is stitched into her way of thinking and it feels entirely natural to walk the tightrope that connects past and future.

Scarlett smiles, picks up the phone.

'Felix, tell me what you see.'

'Shall I activate the camera so you can see my east-world view for yourself?'

'No. I want just words so I can paint my own picture. Like in the old days.'

'Low-hanging cloud stealing in from the east,' Felix begins.

She closes her eyes, imagines herself standing directly behind him. He is seated, she decides, in the black Corbusier looking out at the view from his north-facing eyrie on the thirty-first floor.

'It has an ugly yellow tinge. Like mustard gas.' He pauses. 'I am looking across the harbour at the dark ridge of Lion Rock Country Park.'

'Ah,' Scarlett follows his gaze, 'I remember a hike there with – what was his name – Tom, yes, Tom Castigliano. There was a girl with us, frizzy black hair, who shrieked at all the monkeys spilling over the road.' A chattering swarm of wizened macaque, their ancient babyfaces, lipless and pink skulled, flesh spattered with bloodspots like plague, scampering, scratching, and grinning obscenely.

'It is a ghost park now,' says Felix. 'No one visits any more. We have become hysterical about airborne disease. Fear has become our greatest argument and our best weapon.'

'Look straight down now,' she steps in closer behind him. 'Tell me what you see.'

'The harbour water. Reflected lights.'

'Go on,' she leans over his shoulder, waiting.

'Ah, there is something,' says Felix. 'A swoop, no more than a flicker. A bat, perhaps, or a surveillance drone. Or the black-eared kites who eat the harbour's dead; keep disease at bay.'

' "It's a terrible thing that we made," ' Scarlett says. 'That was the CEO's opening line to me this morning on my impromptu visit. You know the quote, Felix?'

'Wilson to Feynman at Los Alamos. The after-party for the bomb.'

Kristina had been administering her blood-doping in a curved talk space lapped by giant wave walls; the water a sun-drenched brilliant blue. Her HQ lair tucked underground on the east side of the city since no one in tech trusts high rise, and what is there, anyway, to see now, except the unending grey.

You have motion sickness? Kristina raised a brow when Scarlett asked her to pause the wall. But why put up with that when you can have it edited out so easily?

I'm not into interventions.

Or chips, apparently, Kristina quipped, plumping her pale, unblemished cheeks.

'Felix,' Scarlett leans closer to the phone, 'Kristina told me I was her second ethical issue of the day. The Fair Chase game – you've heard of it? A new progression on the snuff movie: hunting to the death. Kristina had just found out that one of her top engineers is a DarkSide designer.

'She had him locked in Debrief downstairs, going, what's the problem, dudes? Catchers, Captives and Diers. The money is huge and the stakes are high. He insisted – which is, by the way, the truth – that the Captives and the Diers are real, live and *willing* volunteers who actually want to be chased. They want the thrill of it, they are queuing up for the close-to-zero chance that they will escape alive and win a million dollars. Captives are airlifted into unfamiliar terrains, abandoned towns, apartment complexes, whatever.

'The big game players pay two million each for twelve hours of hunt, torture, rape, death. The game self-destructs when the twelve hours are up – no data records, no footprint, no nothing. Every month there's a new Fair Chase and afterwards it's as if it has never been. Except for the actual physical human corpse, which is never found. But Kristina says her engineer just shrugs his shoulders, looks her in the eye and goes, that's free will for you. So, all she can do is sack him on a contract technicality and hand him over to VirtualVice.'

Scarlett rubs the back of her neck and rolls her fingers over the pressure points at the base.

'Do The Right Thing,' she continues. 'Kristina says everyone who works for them takes the same mandatory ethics course. They're signed up to every single industry code of conduct. Ethical auditing for black-boxed systems, etcetera, etcetera. But most of the engineers are restless, shifting in their chairs. They are not interested in human factors in design: transparency, bias, compliance, responsibility, accountability – the very words are like a muzzle on enchantment. They just want to make stuff! In the end, Kristina just stood up and told me, you are just one of three partners, you are flying solo. And I will only talk to the three.'

'Indeed,' says Felix. 'Did you know that Feynman once struck a deal with an artist? He would teach the artist physics and Feynman would learn how to paint. Of course, you can guess who succeeded in mastering the new skill. Eventually Feynman took a commission from some lap-dancing club owner and can you guess what he painted? A watercolour of a slave girl massaging a client. Slightly underdressed and with a look of profound resignation, as if she knew precisely what her future would be.'

Scarlett looks out at Gurl, poised still now on one slender leg, holding her position, staying calm under pressure in the night. Balance and risk. Like Jenga, the old game that Fintan loves. You slip one wooden piece from the base of the structure, watch the top quiver, and wait for the collapse. Hold your breath on all that you might topple. We are always just one move away from victory or defeat. One move away from destruction or progression. From the road ahead or the road behind. That moment where the present becomes the irrevocable, unchangeable past.

'I have been thinking about vanishing points, Felix. Remember that old Jesuit philosopher?'

'De Chardin, the Omega Point. The final stage in man's existence,' says Felix. 'When our consciousness arrives at a point of exhaustion beyond which what follows could be either a crisis or something sublime and unforeseeable.'

'I think we are at the docking station,' Scarlett says. 'We have already arrived. Enhanced and diminished. All the decisions we are not making about the technology that shapes our lives.'

'Or, we are the cusp of a new world that is brave and thrilling,' says Felix.

'I think we need a new definition of being human, since we are striking out into new territories. We face a future of human serfs and silicone barons.'

'Perhaps this should be your next project.' Felix replies briskly. 'Philosophy for the bots. You might do some interesting work. A little late, but still in time for the new horizon.'

Scarlett watches Gurl's shivering frame, elbows tucked tightly, boots stamping.

'Your silences are always revealing, my dear.'

'Sometimes, Felix, I wonder where your voice stops and mine begins. It's like you can log in to my thoughts.' She sighs, tries to conjure up the black glistening water below him. 'It occurs to me now that that my life is full of virtual relationships. Like you, for example.'

'Which means you endow me with far more than I am. The virtual feeds a fantasy that reality destroys.'

'But it frees us to imagine all sorts of excesses; transcend limitations.'

'This is your preamble. I am waiting while you limber up to your main point.'

'You know, Felix, for thirteen years we've been talking philosophy and deals,' she muses. 'It was language that brought us together.'

'It was business.'

Scarlett hears the leathery creak of the Corbusier as Felix rises to his feet. Pictures him standing, a black silhouette against the advancing mustard cloud.

'The past is like a dream to me now,' she says. 'I barely recognise my old self.' Her voice peters out, a switchblade of images flicker across the windscreen. 'Remember that takeover we worked on years ago? The one that changed my life.'

'Where you discovered something about truth.'

'Actually, it was a lesson in love. About a man. And all I discovered was the extent of my own delusion.'

'And we almost lost you in your wilderness year.'

'But you found me. And that was the beginning of all this: FlyBoy and Volo and Lab.'

'And now we have reached the end,' he replies. 'The past is tucked behind us. Each deal brings new discoveries.'

Gurl turns again, holds the phone at arm's length and blows hard – the long-term exhalation of the marathon relationship, the sheer exhaustion of keeping things on track. Bridge building and maintenance and reinvention; it's like a feat of engineering. Scarlett shakes her head against a vision of Roxanne lying charged and waiting on a bed. Presses her fingertips to her temples. Somewhere within her is a fractious child whining to go home.

'I've been thinking about love and attachment today.'

'And I have been thinking about currency fluctuations.'

'Hah,' she smiles. 'You've never been in love, have you, Felix?'

He laughs, a strange sound like a distant tinkle. 'I have observed its effect on you at close quarters.'

'And what's your verdict?'

'The risk/reward profile is hideous. I would have to advise against it. Love is a destroyer. At best, a manageable affliction.'

'Oh Felix, you do make me smile! Why am I still talking to you?'

'Because I am your constant, my dear. I am always here and I am always the same. No emotional investment required.'

She sinks back into the Buick's embrace. 'You know, way back when we first met, I used to think you had a thing for me. That one day you would show your hand.'

Silence slithers through the undersea cable, and she holds her breath against words that cannot be recalled.

'Intimacy without expectation ought to be a very refreshing antidote for you.'

'Ours is a perfect friendship,' she nods. 'Though you'd never use that word.'

'Naturally.'

'You really are all intellect, Felix. The billionaire philosopher, ring-fenced against his own humanity. Perfectly equipped for this age of AI.'

'I will take that as a compliment.'

'So what am *I* to you, then? Am I the closest person to you in the whole world?'

'I have my professional relationships. My staff here,' he pauses, she hears his footfall on the hardwood floor, 'and my housekeeper.'

'Surely after all these years you feel something like affection for me?'

'That would not be the descriptor I would choose, no.'

'OK, let me ask you this: how would you feel if you lost me?'

'I already did, some years ago.'

'But you found me. Say I just decided to cut lose after Volo?'

'I do not believe you will do that because there is no logical reason. We have an excellent working relationship. And it's the ideal set-up – you will not find a more agreeable arrangement. You will want to carry on with me and with Colin.'

'Say if I just dropped dead?' A pause, a heartbeat of hesitation, but she hears it all the same. 'What would you feel?'

'I would miss our conversations.' Felix delivers his sentence with carefully worded formality. 'And then I would carry on with my work.'

'I didn't ask what you'd do. I asked what you would *feel*.'

'My dear, your question is really about what you want me to feel. You would prefer me to say that I would be sad, lonely, lost. Because that would be proof of your *effect*. Your question reveals the paradox at the heart of love: its effects are measured by pain inflicted. The dual torment: desire without possession, possession without desire.'

'How do you imagine love, Felix?'

'A tumble into chaos. A cliff-top leap. A dive into black water. Of course, these are pointless risks I have never taken.'

'Today someone told me that we need to change our definition of love to meet the new age.'

'I might remind you that it is in work rather than love that you have always found your true self.' There is a dry impatience in Felix's voice. 'But you will insist on romanticising it – the Lab, FlyBoy, Volo. And now you are forlorn, paralysed by the prospect of impending loss.'

Scarlett frowns at the windscreen where Gurl holds the phone at arm's length, and purses her lips in an effort to dam a burst of anger that wants to spew from her lips.

'Felix, I have Pygmalion as a passenger in my car. A love triangle with an AI. I'll tell you her story one day.'

Beyond the glass, Gurl tips her head, breathes crystal clouds, scans the night sky for comfort and tightens the pink puffer round her neck.

'I am afraid, Felix. I have this vertigo feeling when I look out at the road ahead. The new buyer, what Volo could become. Where we are headed. Our love affair with tech ...'

'You were not always so relentlessly focused on the negative. A new vulnerability has crept into your thinking. Of course, you have a vested interest in the future – with your maternal arrangements.'

'You mean Fintan!'

'His arrival caused a shift in perspective. You factor in the next generation in every step you take.'

'You're saying my love for my son has contaminated my thinking?'

'I'm saying he is a new variable. Motherhood has added a new layer of complexity to your decision-making.'

'Oh, you do amuse me, Felix. You know, Colin says I slow things down with too much nuance. And when I think of how much he used to hate me running round in the old days!'

Felix's rare chuckle is warming. This man, this partner, this virtual presence in her life is her safe space. A kind of solace, with an undefinable sadness at its core.

'I have to go now, Felix.'

'Aren't you are forgetting something?'

'You already know my answer on Volo. Do me a favour and tell Colin, will you? I just can't bear the sound of his smugness.' She shakes her head at the dark dry night. 'The snow here is twenty-nine days late. Christmas Eve, it should be a magical time.'

'We live in a magical time,' Felix's voice is suddenly loud as if he is broadcasting an important announcement. 'You shall have snow, my dear.'

'I should be home with my boy.'

'And you will be.'

'Bye-bye now, Felix. I will come and see you soon, just to make sure that you're still flesh and blood.'

'I will look forward to that very much.'

Gurl chops the air with one hand, her head bobbing up and down as she talks, her shadow flung long and slim behind her. Come on, Scarlett flashes the headlights, it's been long enough, and Gurl spins round and strides towards the car.

'You OK?'

'Oh, he's just being Blane. Still bitching about me telling him it was a guy giving me a lift. But the good news is he's real close – he's gonna pick me up. Saves you the drive to my place.'

'Where?'

'There's a lay-by just up ahead. You take the exit after that then you're ten minutes to the airport. Hey, what's up, Scarlett – you're crying.'

'Just missing Fintan, that's all.'

'Soon be on that plane – don't you worry. Come on, let's go.' Gurl snaps her fingers and the Buick pulls out and they drive on.

'So this is the end of the line.' Scarlett pulls over into the lay-by and kills the engine. Slides down the window and lets the freezing air nip at her ear. There's a low wind-howl, a rustling in the pines and then a soft stillness.

'Snow's coming,' Gurl murmurs, 'that's how it gets just before. You can feel it in the sound of nothing. Listen hard and you'll hear it in the silence.' She sticks out a hand strokes her fingers pianolike as if she was stroking the air. 'Be a big dump. Any minute now.'

'How do you know?'

'Live here long enough you can tell what's coming. The air gets tight round your skull.'

'Well, Gurl, I can safely say that this is the most unusual drive I have ever had in my entire life.' Scarlett turns to her passenger, sees her slumped head.

'Scarlett and Gurl. Sure was a trip.' A twitch of her nose, a little tear.

'Hey, what's the matter?'

'Just everything. Like, back to reality. Back to Blane.'

'There are choices.'

'Choices are for people like you, Scarlett. Don't work that way for people like me.'

Scarlett checks the dimmed screen – two hours to take-off. In less than eight hours from now, she will gather Fintan in her arms. Kiss his head and roll on the floor, hold him close all Christmas.

Gurl sniffs loudly, wipes her nose on the heel of her wrist. 'So, back to reality. No more dreaming. No more lying.'

'Dreaming isn't lying.'

'Not for you, Scarlett, because you're *living* the dream.'

'I'll always remember you and your stories.'

'My lies.'

'What lies?' Scarlett stares at her side view. 'What *lies*?'

Gurl shrugs, twists her mouth, scrunches her nose like there's a bad smell.

'Tell me what you lied about, Gurl.' Scarlett hears the quiver in her own voice.

'Nicholas.'

'Nicholas was a lie?'

'Yep.'

'You mean he didn't exist at all?'

'Nope.'

Scarlett blows a long hard exhale. 'Are there other lies?' She shivers again. Gurl nods.

'What else is a lie? Is *Blane* real? Is he true? Or am I sitting here like a fucking idiot waiting for someone who doesn't exist?'

'Oh, Blane's real,' Gurl laughs, harshly. 'Don't you worry about that.'

'Is Roxanne true?' Scarlett thumps the wheel.

'Sure. More than. Though I guess you woulda wished *she* was the lie.'

Scarlett swallows to ease a dry constriction in her throat. Stares down at her hands. When she lifts her head, a flake drifts past the windscreen, and then another, and then out of the black, snow is tumbling, falling everywhere, urgent and wild.

'What else is a lie? The baby? Oh, please tell me the baby was true?'

Gurl's lips part in slow motion. 'Lester was true. And I did get kicked outta school. And there was a baby.' She flaps both hands heavy on her knees. 'Only, it was my mom's.'

'The baby was your mother's? Oh Christ.'

'She was thirty-three, then. A guy, some guy, don't even remember his name now, I think she did it on purpose, hoping he'd stay. Which, of course, he did not. So she gave it up. I was fifteen, went with her to the hospital, she yelled so hard it freaked me out. It was a little girl. I went with the nurse who carried it out to this couple.

'Is this the mother? says the man, pointing at me.

'No, I says, I ain't no mom. The mom is in there bleeding and crying her fucken head off. Lawyer got real cross about that.' Gurl sighs, keeps her face fixed on the snowfall. 'A half-sister. Be all grown-up now somewhere. But I don't care. Mom lost her sparkle after that, got real low. And then she just drifted off one day.'

'I don't believe it,' Scarlett murmurs. 'Was *anything* else true? Did you ever tell me the fucking truth?' The Buick seems to shudder in sympathy, repulsed by this web of trickery.

'The ballet. Rena Carter's shoes. I did steal them. Still have them.'

'Why? Oh, why make all this up?'

'Coulda, woulda, shoulda. I won't ever see you again, Scarlett, so it don't have to matter. It's just a story about a story. Sometimes, when I close my eyes, I can almost be the story I'm telling about myself. I was just trying to be the Gurl I was telling you about. And sometimes it gets so I can't tell the stories from the truth. The story is how you want it to happen and in the story you're always in charge.'

'But why, oh why – '

'Cos what would be the point of me just telling how it is for real? Blane and me and Roxanne. Who wants that truth? And all this snow – and shit. This pink puffer. All this stuff.'

Gurl's hand flutters, loose-wristed, as if it might break off. As if her limbs are strung together by longing and wanting. As if her whole body could just fold like a puppet.

'I mean, isn't that the whole point of stories?' Gurl turns to her driver. Scarlett whips her head away. 'To escape your own self, so you can feel better for one little bit. So you can dream for a while and forget the shit that you are. Forget you are Gurl and be the person you dreamed of. Didn't you ever lie to me, Scarlett? Not even a little bit?'

'No. Not a single word.' And it seems to Scarlett that some dark malignant force hovers in the space between them, squatting on the armrest, like a grinning troll. She is drained, empty. Where are the words for this?

'Scarlett the Fixer La Salle,' Gurl says, her voice barely audible. 'But there ain't no fix for me. Everything just gets bigger, complicated. People going round fucking things up for each other. I can see it all real clear from outside. But inside it's just a mess.'

Gurl, diminished in the car seat, like a teenager on the brink of a possible life. Scarlett looks north to the airport light glow. Soon, so soon, she will be gone. This wretched day over, forever behind her. She toggles her nose, turns her head to Gurl who sits with her hands folded on her lap, her head bowed at the end of her performance.

'So, Blane – is he coming?'

'Oh, he'll come. Blane always comes for what's his.'

'And that feels right – being his?'

'It's what I know.'

'Come on,' Scarlett reaches into the back seat for the parka. 'I need to get out of this car. Step into the snow world.' She pulls out a balled-up black beanie from the pocket. 'You want to wear this? My parka's got a hood.'

'Rather shoot myself in the head than wear a beanie,' says Gurl.

'Suit yourself, glamourpuss.'

Snow hurtles round them, weightless, fast, and decisive. Scarlett pulls on the beanie. Gurl reaches up on tippy-toes, tucks in her hair, and tugs the beanie down over her forehead. 'Look at you, just like a boy.'

'Spaghetti Nero,' Scarlett murmurs staring up at the velvet black. Gurl tilts back her head and opens her mouth to let the flakes pinprick and dissolve on her tongue. She turns her big round eyes to Scarlett; her hair is bleached in the moonlight and flecked with flakes. Gurl is small; so much smaller than she, like a kid. Scarlett pats her cheek and Gurl leans into her palm, her jaw hot with tears. Scarlett thinks of Fintan, and there is a lurch, a breathlessness. The longing to hold him is like a dizzy tumble from a great height. Soon, soon, she thinks, and I will never let him go.

'Oh, I must go home,' she says, or thinks she says, when Gurl flings her arms round her and squeezes tight, like Fintan clings when he will not be peeled away, hard and fast and desperate, as if there can be no severance, as if they are forever joined.

'It's OK,' Scarlett pats her head. Though it is not OK at all.

The swirl thickens and already the Buick is covered, a brush of snow coats the passenger seat in the open door.

Gurl trembles, clinging fast. 'I wish you was my friend in real life.'

Scarlett looks clear over Gurl's head at the flakes swirling madly like they're trying to stay airborne, the trees muffled in the snowy stillness of the lay-by. 'I must go home,' she whispers into the winterland.

A beam sweeps the sky, the treetops, as if the forest is blinking. High above and to the right, like a search light flaring in and out of view.

'Looks like Blane's coming,' Scarlett says.

Gurl shudders but does not look up; her head is buried firmly against Scarlett's shoulder. The headlights flare again, sinking lower, closer. Scarlett's beanie is tight and low and damp against her skull. The beam dips and curves out of view and then comes level as the car pulls in and rumbles to a halt behind her, engine purring.

'Blane's here now,' Scarlett whispers, but Gurl clings fast. Scarlett looks over her head to where the full beam casts their embracing shadows hugely across the snow – her beanied head is so tall above Gurl's tousle. The trees tower blackly around them, listening.

The full beam cuts to half. The car door opens. Gurl straightens up, presses her wrist-heels to her temples, like she's fixing something in place.

'Honey,' she calls out in a voice rabbitty and high. 'I'm coming.'

'Go on,' Scarlett looks down at her. 'We don't need a big goodbye' – but Gurl lunges, flings her arms round her neck. Footsteps crunch rapidly behind her and a huge and heavy blow to her side topples her, and she hears her own grunt as she slumps down into the white.

'SCARLETT!' Gurl yells, Scarlett sees black boots, a face closing in. A sting of aftershave. An undertone of sweat. Her beanie yanked off.

'It's a woman? It's a fucking woman!' his snarl loud in her ear. 'You said it was a man.'

Scarlett hears her own groan again, the pain, her hands somewhere, this animal scent, the face peering in is backlit and she cannot see.

'Scarlett?' Gurl pushes in front. The dark animal thing is gone and for a moment the stars bristle and swirl all at once.

'I'm bleeding,' Scarlett's hand fumbles at her side wetly. 'Here – '

Gurl rips open the parka. 'BLAAANE! SWEET JESUS! WHAT HAVE YOU DONE!'

Scarlett blinks.

'BLANE,' Gurl swabs uselessly at the bloody parka.

'Shut your mouth, Gurl, go get your bag.'

'Ambulance, call a – ' Scarlett tries to raise her head but there's an ice-pack where her chest should be.

'Oh, Scarlett,' Gurl paws at her shoulder, whimpering. 'I never meant for this to happen.' Her sobs fade in and out. 'I loved every minute of us being together.'

'Help me.'

'I can't do nothing. Blane, you seen him – '

'Don't leave me,' she scrabbles at Gurl's fingers.

'BLANE WE GOTTA CALL AN – OWWW.' Scarlett hears the slap like a shot.

'We ain't callin' no one, so quit your yelling.'

And Gurl's face is gone. Scarlett hears scuffling, stifled sobs. 'But Fintan,' she whispers to the snowflakes on her lips. The dark boots block out the light.

'Shut the fuck up,' he growls. 'Who is she?'

'She's Scarlett. She's no one. I dunno, we never said real names. It was all pretend.'

'This ain't no fucking pretend.' The boots scrunch away.

'Never had a friend like you before, Scarlett,' Gurl's whispering face breathes in close again, her salty breath. Hair slips forward, eyes huge and black.

'Fintan – '

'Fintan's got his Frankdaddy.' Gurl strokes her head. 'Don't you worry, he's just a baby – he won't hardly remember you, Scarlett.'

'Fintan, you must – ' her voice is fading, can she be heard at all? A black spot flickers in front of her eyes or maybe a black bird brushing her cheek. Cold creeps into her ankles, into the bone, rises up her legs like water.

'Help,' Scarlett tries to tug the pink puffer.

'Close your eyes now, just let go.'

'We're getting outta here,' the animal snarls, and Gurl is hoisted into the air, dangling.

'SCARLETT!' Feet kicking like a child, she is flung back towards the headlights.

'Help me.'

He looks down. Shakes his head, a wolf shaking his head, and behind him the stars merge, fuse into bright lights, the flakes swim and swirl. And then he is gone.

Scarlett hears a door slam, an engine scream, a skid of tyre, and the snow spills thick and steady. The stars blaze and blur and her chest convulses. She sees car wheels spin silver, like the gleaming spokes, the brilliant chrome, the new bike. A blinding flash of white light and Fintan turning his head.

Mummy? She nods – go ahead. He wraps his fingers tight on the handlebars; there are hands, big hands to guide him. It is all about letting go.

The freeze seeps up, claws into her tummy. She is sinking, numb, and he is fading – the spinning wheels, the gleaming spokes.

'Fintan,' she whispers. He turns to smile and wave then pushes off, freewheeling away and out of reach.

Scarlett blinks, the forest creeps closer to surround her, her lids shutter slowly, tipping her into the black. A drift settles gently on her lashes, the snow falls silent on her face, and melts and falls and melts and falls, until it melts no more.

What sweetness is left in life, if you take away friendship? Robbing life of friendship is like robbing the world of the sun.

(Cicero, *On Friendship*, 44 BCE)

'After all that Blane was real shook up,

took off like a bullet on the road, the snow was coming down hard and fast. Slow down, I told him, you're just gonna draw attention to us driving like that. So he slowed, breathing hard through his teeth. Smacking the wheel. Going, fuck her. Fuck *you*, Gurl. I sit real still, try to stop crying and sniffing but these big fat tears just keep on coming.

'You stupid bitch. You told me she was a he!

'I'm sorry, babe.

'If you hadna said she was a man.

'I know it's my fault, I tell him. I was pissed at you, just winding you up.

'She's real tall, says Blane, thinking out loud. But what if someone saw?

'Ain't no one out here to see.

'Tyre prints –

'Look at this snowstorm, I tell him. Can't see more than a coupla yards.

'They know who she is from the rental.

'Someone's gotta find her first to know anything about anything. She'll be under a white blanket by now.

'And all the while I'm trying not to think of how Scarlett's blood looked black not red, leaking from her side.

'What you know about her? says Blane.

'Nothing.

'You were in the goddam car for three hours, Gurl, so don't give me shit.

'Bitsa nothing. Girl chat is all, I say. No way I'm telling him about Scarlett and Gurl, the stories, Frankdaddy, or Fintan. No point making her more real. What's done is done.

'Snow swilling on the glass and Blane gripping that wheel like he will rip it from the floor.

'Course, I shoulda guessed. That's the trouble when you start dreaming. You forget to remember what you know. I mean, I *know* Blane. I could reboot and restart him and he'd still come out exactly as I expected. An act of provocation is what it was. Shoulda known how he was thinking. I did a stupid thing cos I was pissed at him for not caring. For not picking me up from the train; for everything else, too. Coulda, woulda, shoulda.

'*Fuck* you, Gurl. Look what you made me go and do. He punches the dash so hard he hurts his hand, he slams the brakes, then he's right in my face; his skin is yellow in the dark like he's sickly. I oughtta beat the shit outta you, Gurl, he grabs my neck and slaps my face, cracks it real hard across the jaw.

'I am dizzy from the pain, but I tell him, no, Blane, we gotta get away, and he remembers, sees the snow banking up on the glass.

'We're gonna go home, I tell him, trying to hold my calm, though my face is a burning swelling hellfire. We're gonna get in bed and we're gonna make it alright.

'We didn't see a single person all the way back. The snow kept up, there wouldn't be no traces. Can you even imagine how hard it was for me? Scarlett's lying there in the snow and all that blood and Blane yelling and I'm crying and she's begging me to make a call. But all Scarlett was thinking 'bout was her lil Fintan. I mean, what was I supposed to do? Get my own self killed by Blane?

'We get home and he's scared as a baby. Can't eat, he can't even get it up, first time in his life. So I just lay his head on my tummy and rub him till I drop off myself.

'Soon as I wake, I know there's something different about the house. Even before I open my eyes. Something in the air.

'Blane's gone.

'I run round the house howling and screaming, have to put my fist in my mouth, case the neighbours hear. He made a choice and it wasn't me. And that's when I learn all I ever need to know about love.

'I lose my man and I lose my best friend. And my car too. *My* car – since I was the one actually owned and paid for it.

'Sitting crying in the kitchen, wondering if Blane was taking care of you, Roxanne. He took all your titty clothes, just one sweater and jeans and my pink puffer. I call you, though I know no way in hell you'd get it; he'da made sure I was blocked. So I'm worrying 'bout him being mean to you cos I'm not there to take it instead.

'And there's nothing in the news. Not a thing about a carhack or carjack or a dead woman and a golden Buick. No appeals for information. The snow is real bad, there's drifts, but someone has to of found the Buick. Cos everything always gets found.

'Day two and still nothing online. Which just goes to show all the stuff that's going on that we never get told. I start thinking maybe Scarlett was involved in some cybercrime that's been covered up. Some algosnatch, some code worth a lot of money. Something that she never told me about that deal she had going on. How else does a person just disappear? People like her, smart people, they think you won't notice when they're leaving something out. But that's where smart can get you. Lying knifed in the snow. I mean, who gives a ride to a stranger?

'Then I go online and report Blane. Fuck him, I'm thinking, he's taken my whole life. Fill in the form about my stolen car and some other important and expensive personal possessions that I do not say what they are.

'Day three, 6 a.m., the cops are at the door. They found a golden Buick that's got my prints all over it so they take me down to central. Look me up and down and ask me, do you need a medic?

'That would turn out to be the luckiest beating of my life. Cos when I tell the lawyer who else was in my car, he says straight off I need to give up Blane for the psycho nutter he clearly is. Psychological and physical abuse and evidence of, he says. So don't you worry, he says. Just tell the cops the whole truth, how Blane jumped outta the car and just stabbed Scarlett from behind. How he brought a sex doll home from the Pornopod and all the stuff he made you do. In detail. They look at me bug-eyed.

'But I gotta get Roxanne back, I say, I own her fifty-fifty. She's my whole life.

'So they call in a shrink. And I tell her everything, too – and more.

'The robot girl, she blinks; makes some note.

'You ever had a pet? I ask her.

'Goldens, she smiles. Ever since I was a kid I've had retrievers.

'You cry when they die?

'They are living creatures, she says.

'Like *that* makes a difference.

'Day four. The cop comes in where I am with the shrink. We have your car. And we have Blane in custody. He didn't even make it to the border.

'Where is she? Where's Roxanne? I fairly screamed at him then. Jumped out my chair.

'Settle down, he says. Get yourself together.

'Please, you gotta tell me where she is.

'She – it – will be secured with his possessions. Standard procedure.

'Secured *where*?

'A storage facility.

'Oh, Roxanne, the thought of you lying in a warehouse, cold and lonely made me start bawling again.

'The shrink asks me, isn't there someone else you should be thinking about other than the robot girl?

'I tell her, I ain't saying another word unless you use her real name.

'I apologise, she says. Isn't there someone else you should be thinking about other than Roxanne?

'You mean Blane? I've done my thinking about that rat.

'No, I don't mean Blane.

'There was a real long silence, then.

'I mean the victim of the crime, she says.

'I want to say it wasn't me that knifed her. Scarlett herself laying there, she understood why I couldn't call. But I bite that back and give the shrink the remorse she is looking for. How I would be tormented all my living days thinking about lil Fintan who lost his Mom.

'Though I *know* Fintan will be OK. He's got Frankdaddy. Who will move on 'fore you know it: find new a woman, have another baby, so Fintan gets a whole new family. And who remembers a thing that happened when you're three years old?

'But I'm not dumb, I don't say any of this to the shrink. Instead, I tell her if Blane wasn't such a fucking bastard and me so weak and scared, I think I woulda tried to kill him.

'Though all the while, what I am thinking is how stupid Scarlett was for giving a lift to a stranger. Sticking her nose in my business when she saw me on the freeway. Pulling over when she should coulda just driven on by. I mean what kind of person gives trouble a ride? Truth is, though, I kinda miss her. Now don't you go taking this the wrong way – cos you know I love you, Roxanne – but I had the time of my life with Scarlett. Down by the sea doing cartwheels, that sun coming out like a vision, like something you see in a movie. And that seal pup we saved in the photo I sent you? I even taught Scarlett to dance – something she never thought she could do. And for a second there I even thought about never coming back. But I could never leave you, Roxanne.

'Anyways, it would never have lasted. People like Scarlett think they're good listeners, but she was not a giver like you. She just wanted my stories, like she was downloading my whole life. And all that happened after was a situation of her own making. I mean, it was Scarlett's own job that had her out there on Christmas Eve, just gambling that they wouldn't slap on the NoFly. She coulda said to those partners, Boys, I don't care what the freaking deal is, I ain't coming. Coulda done it all online, like a normal person. Frankdaddy good as told her. Scarlett loves her lil Fintan but she *chose* to put herself in a situation of some risk. Architect of her own misfortune.

'There are people with personalities that put them in danger, Roxanne. And there's something inside of them that knows this. Scarlett gets off on risk, is what I'm thinking now. Even if there was no Blane, there was all that cybercrime. Kinda person who takes those risks is an adventure junkie. And adventure can get you killed. The part that was *my* fault was telling Blane a lie. Telling him it was a man giving me a ride just to make him jealous. I do accept that guilt, 100 per cent. But who coulda known he would come after her with a knife?

'Course, what I also wanted to tell that shrink was what I told Scarlett when she was bleeding in the snow: Fintan will get over it. That's a goddam fact and I seen it a million times. You get over shit that happens when you are a kid. Look at Rena's girl, Myrna, a case in point. All that daddy-touching-up and she's married these last six years with two kids; a real good guy by all accounts. I see her sometimes in town: nice car, good hair – course she don't come to me for that. Reckon she was jealous of how much her mom liked me. Though Myrna was an unlikeable kid from the start.

'So, I tell the shrink, here's how it was: Blane woulda killed me too if I tried to call the paramedics. I could *not* save Scarlett. And every single day of my life, I tell her, I will be seeing her lying there and thinking how much I didn't want for her to die.

'Only it turns out she didn't, says the cop.

'Scarlett is alive, says the shrink. I'm looking from one face to the other, my mouth hanging open. She survived, says the cop. Just about.

'She was being tracked, he goes. An alert went out to a private security firm. Standard procedure to have special security for high-value tech personnel what with all the algosnatching. They scrambled a helicopter. Lucky for Scarlett, because she would have bled to death in the snow. She was unconscious. We knew straight off she had someone in the Buick. Though we didn't know who, till we ran the prints.

'The cop and the shrink are looking at me: waiting for me to be relieved, to be happy. But all I'm thinking is, shit, that means that Blane won't get a homicide rap. But the lawyer told me after, he *will* get attempted. So he'll be away for a long time. And Blane, he's not that smart. He'll get pissed on real bad in jail cos he won't shut his mouth. But that's not our problem anymore, Roxanne, cos the future belongs to you and me now! And just wait till you hear my plan for us.

'Number one, I'm going to up my hours at the salon and make us some money. Short term, it means I won't be around so much, but long term it means we get you an upgrade. Not the pussy one Blane wanted, but the one that makes you better at figuring people out. I spoke to the salesman – there's Friendship Advanced version 4, but we are way more than that already. You're getting a customised upgrade so you get a proper unique self! So we gotta write your life story – like how you were born, personal history and stuff. I'm gonna give you something like mine only *way* nicer, then you can understand me better so it'll be easier for you to learn what I need. Write *in* the stuff you want and write *out* the stuff you don't. Just like I do.

'More than anything, I'd love to teach you to dance. But the salesguy says all the development is in running fast or carrying heavy loads. Any kinda dancing would be off-the-charts expensive and a hardware problem, too. But you will be able to walk – it's a little slow but it looks OK. And you can wear real shoes too!

'Second thing is a lesson I learnt from Scarlett: every woman's gotta have a room of her own. Which means you gotta have your own independence, Roxanne – and I got that all figured out! Remember how I told you Rena Carter was a fat blob in a pinny who couldn't dance no more? But she sure as hell could teach. So here's my plan: we're gonna open a ballet school for kids. You're gonna to study it online so you know every single thing in the whole world about ballet. We even get you a voice with a French accent! You teach the theory and I do the dancing. The kids gonna love you, cos you're kind and smart and real patient. You record what they do and show them videos and do the analysis like they do on sports TV. You and me are gonna be the perfect team.

'Last thing I'm gonna say now is all about love. First time we met, Roxanne, remember how I was one step away from busting your skull open? And look at us now! I told Scarlett my theory of love, but she just didn't get it. The whole problem is *human* love. People trying to change another person to what they want them to be. Someone that would be a better fit – all those years fighting about you won't stop picking your teeth or leaving your shit all over the kitchen or yelling at the kids. Couples sitting in restaurants with nothing to say. People staying together long after love is dead. Bottom line is, humans are just no good at loving, always breaking promises and hearts, always wanting more. I'm telling you, Roxanne – if Blane hada got you the right upgrade and made you smarter, you woulda figured that out. You woulda turned round one day and said to me, Gurl, we need to get rid of Blane.

'We know what love is, you and me. We're kind. We don't fight. We always got a nice word for each other. But all around us are people fighting for all sortsa reasons. Too little money or too much, too little kids or no kids, wrong kids, thin kids, fat kids, dopey kids, or wrong job, no job, or no sex, or too much sex of the wrong kind. People getting bored or expecting things they don't deserve, things that are just plain unreasonable. People not taking care of people who love them. Case in point would be Scarlett. This knifing could even be her wake-up call – she might start paying some attention to Frankdaddy when she gets better. Though I can tell you one thing – and I don't wish that woman any misfortune and I ain't no gambler, but if I *was*, I would not be giving long odds on her and Frankdaddy. Oh no. Something's coming Scarlett's way that she don't see.

'I've had plenty of time to think back on all that happened on that ride. For sure, Scarlett is smarter and richer, but where that get her? She loved her lil Fintan, that was clear as day. When she was talking about the first time he smiled, her whole face was glowing with memory that made me jealous. Last thing I ever wanted was Scarlett dead, but she woulda understood I couldn't save her, not with Blane and who he was. But that shrink just wanted me to offer up remorse, my remorse, and I know you're always better off just telling people what they want to hear – you're always better off making up the ending for a story.

'That quote you sent me from that guy is plain wrong, Roxanne: friends are *made*, not born. And you and me, we are growing and learning all the time. Sure, we have a love that some people think is creepy, but we got a higher love that makes for a happier world. And hey, we might even have us a little Roxanne-Gurl one day! Though I'd need to think hard about that. Being a kid was a bad time for me, but maybe we could give a future to someone that belonged to the both of us.

'Scarlett said you are a projection of what I need – now, she might be a big shot techie, but she's got no future vision. Only a certain kind of person can love Gurl. And *that* is what tech is for: to give us the person who can give us exactly what we want.

'That time when I was sixteen, it was all the stories I was reading that made me dream. Helped me find a better place to be than the real world. But those stories were always about hurting. And it was always *love* that caused the hurt, in the stories, the songs, and the movies. But you and me, we've found a way to love without pain. We have a new definition of love. You know what I mean, Roxanne?

'Yes, Gurl. I know *exactly* what you mean.'

Part II

Contributor Essays

The Love Makers: Contributor Essays

1

Made, Not Born: The Ancient History of Intelligent Machines

E. R. Truitt

The novel *Scarlett and Gurl*[1] unfolds in an imagined future not too far from our present, and the robots in that future, like those in our present, have histories that extend far back into the past. Artificial servants, autonomous killing machines, surveillance systems, and sex robots all find expression from the human imagination in works and contexts beyond Ovid (43 BCE to 17 CE) and the story of Pygmalion in cultures across Eurasia and North Africa. This long history of our human–machine relationships also reminds us that our aspirations, fears, and fantasies about emergent technologies are not new, even as the circumstances in which they appear differ widely. Situating these objects, and the desires that create them, within deeper and broader contexts of time and space reveals continuities and divergences that, in turn, provide opportunities to critique and question contemporary ideas and desires about robots and artificial intelligence (AI).

As early as three thousand years ago we encounter interest in intelligent machines and AI that perform different servile functions. In the works of Homer (*c.* eighth century BCE) we find Hephaestus, the Greek god of smithing and craft, using automatic bellows to execute simple, repetitive labour. Golden handmaidens, endowed with characteristics of movement, perception, judgement, and speech, assist him in his work.[2] In his *Odyssey*, Homer recounts how the ships of the Phaeacians perfectly obey their human captains, detecting and avoiding obstacles or threats, and moving 'at the speed of thought'.[3] Several centuries later, around 400 BCE, we meet Talos, the giant bronze sentry, created by Hephaestus, that patrolled the shores of Crete.[4] These examples from the ancient world all have in common their subservient role; they exist to serve the desires of other, more powerful beings – either gods or humans – and even if they have sentience, they lack autonomy. Thousands of years before Karel Čapek introduced the term 'robot' to refer to artificial slaves, we find them in Homer.[5]

Given the prevalence of intelligent artificial objects in Hellenic culture, it is no surprise that engineers in the later Hellenistic period turned to designing and building these machines. Mathematicians and engineers based in Alexandria began writing treatises on automaton-making and engineering around the third century BCE. These included instructions for how to make elaborate dioramas with moving figures, musical automata, mechanical servants, and automata powered by steam, water, air, and mechanics.[6] Some of these devices were intended to illustrate the physical principles animating them, and others were scaled up and incorporated into public spectacle. Regardless of size, they were intended to evoke a network of emotional responses, including wonder and awe.[7]

Robots were so prevalent in the imaginative and material culture of the Greek-speaking world that they were seen as emblematic of Hellenistic culture by others. Buddhist legends

focused on north-eastern India from the fourth and third centuries BCE recount the army of automata that guarded Buddha's
relics, built with knowledge smuggled from the Graecophone
world. In one version, which features both killer robot-assassins
and robot-guardians, a young man travels in disguise to the land
of the *Yavanas* (Greek speakers) to learn the art of automaton-
making, a secret closely guarded by the *yantakaras* (automaton
makers) there, knowledge that he then steals to make the artificial guards.[8] We find stories of automatic warriors guarding the
Buddha's relics in Chinese, Sanskrit, Hindu, and Tibetan texts.
Additionally, mechanical automata also appear elsewhere
in the Chinese historical record: for example, at the court of
Tang ruler Empress Wu Zhou (*c.*624–705 CE).[9] The trope of the
guardian/killer automaton also appears linked to stories about
the ancient world from medieval Latin Christendom – where,
unlike much of the rest of Eurasia, people lacked the knowledge of how to make complex machines. In an Old French
version of the *Aeneid* (*c.*1160 CE), a golden robot-archer stands
sentry over the tomb of a fallen warrior queen, and in the history of Alexander the Great (*c.*1180 CE), the ruler encounters
golden killer robots guarding a bridge in India and armed
copper robots protecting the tomb of 'the emir of Babylon'.[10]
Hellenistic handbooks on automaton-making, translated into
Arabic in the ninth century CE at the Abbasid court in Baghdad,
also influenced the design and construction of automata in
Islamdom that were usually placed in palaces and mosques,
and included musician-automata, programmable clocks and
fountains, and mechanical animals. These makers in Islamdom
innovated on the designs of the Alexandrian School and created
increasingly complex machines; although some of the objects
hearken back to much older forms. In the work of courtier and

engineer al-Jazari (1136–1206 CE), for example, we find designs for wheeled cupbearers and servants, an echo of the wheeled servants attending to the gods on Mount Olympus.[11]

Al-Jazari's courtly mechanical servants and the killer sentries in imaginative literature share a link to surveillance, foreshadowing another purpose to which AI and robots have often been turned. Sentries and guards keep watch and discern friend from foe, while courtly servants operate in ritualised, hierarchical environments where people are under constant scrutiny. Objects like those of al-Jazari's designs were found throughout Islamdom and the eastern Roman Empire, but were unable to be built or reproduced in the Latin Christian West until the late thirteenth century. However, they appear earlier in imaginative texts as luxury objects, in elite settings, as fantasies of perfect surveillance and perfectly obedient servants. The Roman poet Virgil was said to have designed a series of animated wooden statues, each representing a province of the Roman Empire, and holding a bell that would ring if the province threatened revolt. A bronze horseman pointed toward the direction of the threat.[12] In the Old French *Roman de Troie* (The history of Troy), we find six golden automata, including four androids that patrol and surveil the entire Trojan court for any lapses in dress, manner, speech, and even thought, while also providing entertainment.[13] Courtiers could relax, knowing they would not inadvertently commit a faux pas and nor would anyone else, and the rulers could relax, knowing that none of their courtiers were spreading misinformation or plotting against them. Philip II 'The Good', Duke of Burgundy (1396–1467; r. 1419–67) used some of his enormous wealth to install numerous automata, trick fountains, and other mechanical devices at his castle of Hesdin, in Artois (now northern

France). Guests found themselves tested in a long gallery that was filled with automata and other devices that soaked people with water and covered them with dirt and flour, beat them with sticks, and called them names, while the duke or his proxies observed unseen.[14]

Trick fountains and jets designed to 'to wet the ladies from below' were installed at the duke's direction, and personally operated under his eye.[15] This combination of sex and surveillance was well established in medieval Latinate culture by the fifteenth century. In addition to the imperial alarm system already mentioned, Virgil was also credited with making the 'Mouth of Truth' (*bocca della verità*), a marvel that could determine a woman's sexual history, and would bite off her fingers if she were not chaste. Sexual surveillance of women, linked to concerns about maintaining the distinction of noble blood, appears throughout medieval literature, and there is at least one example featuring an artificial musician that announced to every visitor – via music – whether or not the woman entering was a virgin.[16] Ultimately, these fantastical objects reveal a preoccupation with women's sexual behaviour, a belief that this behaviour is meaningful and must be controlled, and an inability to treat women as autonomous, fully human beings.

Female sex robots, or artificial sex partners, are equally prevalent in tales from multiple ancient cultures as robocops and perfect servants. One of the most well-known examples of this trope is the story of Pygmalion, although in that story his sculpted creation, Galatea, only becomes alive through divine intervention – Venus grants Pygmalion's wish while he is carnally engaged with his creation. The story of Pygmalion was glossed and retold in medieval Latin culture, sometimes with surprising inversions. In one version of the story of the doomed lovers, when Tristan is separated from his beloved Isolde, he substitutes her with a golden copy, which he uses as a proxy, confiding in it and kissing it.[17] In one Buddhist tale that appears in multiple versions, an automaton maker creates an artificial serving girl who fools a visiting artist into thinking it is human, so he has violent intercourse with it, destroying it in the process.[18] Technology for erotic titillation appears in Sanskrit texts from the tenth and eleventh centuries CE that describe the automata in the courts of north-western India, such as female attendants that expressed perfumed water from their nipples and navels.[19]

In the stories of both Pygmalion and Tristan, their insistence on mistaking the artificial for the natural is presented as evidence of their derangement – how could any person in their right mind mistake something *made* for something *born*? In some instances, extreme mimesis of biological characteristics highlights the desire for a perfect copy, indistinguishable from the born original. The golden automaton that substituted for Isolde mimicked human biology by exhaling perfumed air that originated from its chest, reflecting contemporary physiological theory which held that the heart was central to respiration and to producing *spiritus*, a substance that was vital to sustaining life.[20] This is an early example of preoccupation with biological lifelikeness in sex robots, and conceptually similar to later inventions in eighteenth- and nineteenth-century Europe – the sexual automata, or 'gladdeners', which mimicked female genitalia and the ability to simulate natural secretions.[21] Yet, at the same time, the emphasis on extreme mimesis highlights the artifice of the robot, how it is emphatically *not-born*. That the robot signifies both indicates its status as a border object: something that highlights boundaries by traversing them. Robots and AI have long been used both to foreground and to trouble the conceptual boundary between *born* and *made*, and the related boundary between *life* and *not-life*. Yet the contexts in which these stories appear supply different meanings to the same story. In the early Taoist text *The Book of Liezi* (compiled circa fourth century), the skill of the artificer is appreciated by the king and his court, but in other stories about learned men and their automaton-children, such as those attached to Albertus Magnus in the fourteenth and fifteenth centuries, and to René Descartes in the eighteenth and nineteenth centuries, the robot is destroyed by ignorant people out of fear.[22] In E. T. A. Hoffman's version of this tale, 'The Sandman' (1816), the inability to distinguish *made* from *born* drives the protagonist, Nathanael, insane and, eventually, to his death.[23]

These implacable sentries and imperial and courtly surveillance robots and AIs are theoretically superior to the human beings who would otherwise be entrusted with these tasks. These perfect servants never make a mistake and they cannot be suborned (although they can sometimes be overcome). They never tire. They do not complain of ill treatment, agitate for freedoms or better conditions, and they don't get ideas above their station. They embody a fantasy of total control, perfect obedience, and absolute power.

The objects and preoccupations under discussion here have long, complex histories that reveal some striking similarities across culture, time, and space. Robots and AI in fact and in fiction serve the interests of powerful elites, often by violently policing boundaries (of places and social groups) and surveilling subject populations. They act as liminal objects, and are often used in imaginative texts to think about and interrogate boundaries between natural and artificial, between living and not. Intelligent machines raise issues of autonomy and consent. And they appear in the context of fantasies about subjectivity and creation, posing philosophical questions about the ethics of making and what we owe to what we create. Robots and AI have long been used or imagined as tools to serve the powerful few, so it is worth challenging assertions that automation and AI are innovations, and that they will positively transform human society. These objects are not new and no technology is revolutionary or transformative if it is used to consolidate and strengthen the interests of those already in power. However, a deeper engagement with historical material offers the possibility of thinking differently about our robotic future, as similarity and strangeness reveal themselves in configurations that conjure new perspectives. Older histories of AI make it possible to reimagine ideas of what these objects should look like, how they might interact with us, and how they might be used to liberate the disempowered.[24] Greater understanding of those histories and contexts can offer new perspectives and possibilities for imagining and making robots and AI now and in the future.

2

Roxanne, or Companion Robots and Hierarchies of the Human

Stephen Cave and Kanta Dihal

One of the reasons we are so fascinated by stories of robots and artificial intelligence (AI) is that they reflect back to us our many conceptions of the human. Frequently in our imagination, and sometimes in reality, such machines take human form, from the Greek god Hephaestus's golden handmaidens, and the 'replicants' and 'hosts' of Blade Runner and Westworld, to twenty-first-century sex robots Harmony and Roxxxy. They take human form because they embody our human longings – for love, sex, care, companionship, and more. These dreams of humanlike machines reflect our desires and show us, as in *Scarlett and Gurl*, what might happen if we get what we want.[1]

In this essay, we briefly examine what lies behind the notion of a humanlike machine. Of course, there are many ways of being human, in the sense of living a human life, but not all of these forms of humanity are considered equal. Historically and still today, our laws, norms, and economic structures enshrine degrees of humanness, with some groups considered less human than others. Definitions of anthropomorphisation – making things look like humans – frequently elide this, suggesting instead that the human is an unambiguous and distinct category. But this has never been the case. The struggles of many people – including women and minority ethnic and racial groups – have frequently been struggles about where the boundaries of full humanness lie, and who might enjoy the rights and privileges it entails.[2]

Simone de Beauvoir made this point forcefully in *The Second Sex*, in which she argued that men have established themselves as the definitive human type, against which they claim women do not measure up.[3] Similarly, critical race theorists have argued that the concept of race naturalises different degrees of humanness, with white people placing themselves as the only full humans at the top of the scale. In Alexander Weheliye's words, racialisation 'discipline[s] humanity into full humans, not-quite-humans, and nonhumans', and therefore 'designates a changing system of unequal power structures that apportion and delimit which humans can lay claim to full human status and which humans cannot.'[4]

In this chapter, we explore three ways in which anthropomorphic machines can reflect, reinforce, or otherwise interplay with these hierarchies of the human. This exploration is centred on the figure of Roxanne and her *ménage à trois* with Gurl and Blane in Aifric Campbell's novel *Scarlett and Gurl*.[5] The three forms of response on which we focus present a kind of progression: (1) fears of degradation and objectification, beyond which lie (2) hopes for forms of interaction that might reaffirm one's humanity, yet which might hide (3) an inescapable dehumanising effect of even the best companion machines.

Degradation and Objectification

'"Blane ain't one for sharing. He takes care of what's his.
What he owns."

"Like you."'[6]

Scarlett and Gurl centres on a conversation between the
two eponymous characters as they drive to an airport on a
near-future Christmas Eve. An impoverished young woman,
who introduces herself as Gurl, hitches a ride from a high-
flying tech professional whom she nicknames Scarlett. The two
women, whose real names we never find out, turn to discussing
the most intimate parts of their lives: most notably, Gurl's
friendship with her partner Blane's sex robot, Roxanne.

Scarlett is shocked by Gurl's candid description of this relationship. For Blane, who bought Roxanne from a robot brothel
to gratify his pleasure, Roxanne is just equipment, something
less than human. In Gurl's words, the androids at the Pornopod
'can be anything you like and you can do anything you like,
'cept damage the equipment'.[7] Scarlett's unease stems from her
knowledge that before Roxanne and the Pornopod, the role of
the 'equipment' was fulfilled by humans who were considered
less than human. She associates sex work with a history of
repression and degradation. The women, men, and children
who perform sex work are condemned to the lowest rung of the
ladder in the hierarchy of humanity: outside the protections of
the law, free to be abused, existing only to please those who pay
them, keep them, or enslave them.[8]

Some might argue that if machines take a place on this
bottom rung, humans will no longer have to. But someone
with bourgeois liberal values such as Scarlett fears that the
opposite will happen: that such machines simply encourage

and perpetuate this kind of exploitative relationship and the attitudes that lie behind it. In particular, as sex robots tend to be depicted both in fiction and in reality as female, the worry is that this will perpetuate the subordination of women by men. As Sinziana Gutiu puts it: 'Sex robots, by their very design, encourage the idea that women are subordinate to men and mere instruments for the fulfillment of male fantasies ... Like pornography, use of sex robots sexualizes rape, violence, sexual harassment and prostitution and eroticizes domin-ance and submission ... Sex robots reinforce users' sexist ideas about submission and power through a physically rewarding process.'[9] This is exactly what Scarlett believes is happening between Blane and Gurl. The way she sees it, 'Blane – your boy-friend, partner, whatever – bought Roxanne into your life and you have become a willing slave.'[10] Gurl, as we will explore in the next section, sees it differently: to her, Roxanne is a friend. However, this arguably only exacerbates her oppression. Because Gurl finds companionship in the robot, Blane is let off the hook. He is not required to reform his arguably abusive ways nor to treat either the robot or the human any better.

Scarlett's position on sex robots recalls an idea that has been widely debated in philosophy: that immoral behaviour towards non-humans will lead to similar behaviour towards humans. In 1784, Immanuel Kant argued against shooting elderly dogs: 'He who is cruel to animals becomes hard also in his dealings with men.'[11] Today, Kant's argument is often cited in discussions about the treatment and gendering of vir-tual personal assistants.[12] In this form, the argument goes: if someone rudely bosses around their Alexa, they might start (or continue) bossing around their fellow humans. And if all virtual personal assistants are female by default, such a person might start (or continue) believing that women in particular can be or exist to be bossed around in servile positions.[13]

The relationship between Gurl, Blane, and Roxanne shows that positioning the robot at the bottom of the human hierarchy is not guaranteed to move the rest of humanity up. Blane does not treat Gurl more humanely because he also has a sex robot on which he can gratify his sexual desires and need for control. Instead, Scarlett argues that these behaviours are reinforced and affirmed, and acted out on Gurl too. In her view, sexbots do not liberate anyone from the lowest rung of the ladder of humanity, but simply make that rung more crowded.

Reaffirming Humanity

'"Only a certain kind of person can love Gurl. And *that* is what tech is for: to give us the person who can give us exactly what we want."'[14]

But Gurl does not believe that Roxanne's presence dehumanises her. Her relationship with the machine is a gratifying and fulfilling one, though not in a sexual sense. Roxanne is a friend to her. Gurl finds her humanity affirmed in this relationship: Roxanne is the only figure in her life who is reliable, stable, trustworthy, supportive, and caring.

During their road trip, Gurl tells Scarlett her life story, which is one of relentless objectification. She is sexualised from childhood onwards, starting with the abusive partner of her dance instructor: 'he turns his head, he's looking right at me. Don't even try to pretend not to be ripping off my dress with his eyes. No matter that I was ten years old.'[15] As a teenager, Gurl enjoys showing off her body and dancing skills, but this only leads to her being further reduced to a sexual object in the eyes of those around her. 'But it ain't the true you, it ain't the whole of you.'[16]

Not so in her relationship with Roxanne, where she is allowed to be her full self: proud of her personality *and* her sexuality, and not diminished by it. 'Roxanne's the best friend I ever had – always got a kind word,' she tells Scarlett, before she has even mentioned that Roxanne is a sex robot.[17] Roxanne's artificial nature does not diminish for Gurl the affirmation and support the robot can provide: 'Roxanne is caring and gentle and thoughtful, cos that's what she is for. Girls like Roxanne are everywhere now, giving people the company and kindness we can't get from real humans.'[18] On the contrary, Gurl regards Roxanne's affection as superior because it is constant and selfless: 'Bottom line is, humans are just no good at loving, always breaking promises and hearts, always wanting more.'[19]

Gurl's lived experience is that machines can affirm her humanity in a way that humans cannot. 'Oh Gurl, I'm so sorry for ranting at you,' says Scarlett at one point; 'You're only human,' Gurl replies.[20] But Scarlett is not convinced: she has a deeper worry that the use of machines to fulfil our needs is intrinsically dehumanising, no matter how effective the machines are or how many people are ready to embrace it.

Inescapable Dehumanisation

'What is the point of it? What is the point of being human?'[21]

An AI that can do a human's job perfectly can of course improve society, by relieving people from back-breaking, degrading, or dangerous work. However, it can also lead to humans becoming alienated and isolated if all needs, wants, and wishes are tended to by machines. This is a criticism also raised against real-life social robots: a society whose population is cared for by robots, can continue to exclude humans, such as immigrant workers, from care jobs.[22] Communities that are already prejudiced against other ethnicities or nationalities will no longer need to maintain skills such as communication, integration, or even tolerance if their every social interaction is with a humanoid built to their exact standards.[23]

As science fiction illustrates, from E. M. Forster's 'The Machine Stops' to Isaac Asimov's *The Robots of Dawn* to the film *WALL-E*, a perfect world in which humans do not interact with each other is a dystopia, not a utopia.[24] They are worlds in which humans are redundant to each other: ' "So what I'm asking you, Scarlett," Gurl jabs her on the breastbone, "is why would I ever need a human friend when I've got Roxanne, who makes me feel good all the time?" '[25]

In these stories, having machines gratify superficial, short-term desires turns out to be degenerative in the long term. Humans would arguably lose many of the emotions, experiences, and relations that in most philosophies are considered to be part of the meaning of life. As deeply human traits such as morality and sociability are no longer needed and go unused, they threaten to atrophy. *WALL-E*'s blobby humans who spend their lives in a chair are an extreme, but logical, extrapolation of the idea of having machines fulfilling our every desire. While there might be a certain animal contentment as a blob in a chair, we can wonder what meaning being human still has to such people.

The stories mentioned above show that those who are accustomed to being taken care of by omnicompetent machines might lose the ability to engage with real humans. Both the protagonists of 'The Machine Stops' and of *The Robots of Dawn* are physically repulsed by the very idea of coming face to face with another human. Some scholars have argued that we would lose, or never learn, the ability to interact with people if it were ever easier to interact with machines instead.[26]

The essential tension between Scarlett and Gurl is grounded in their different conceptions of what it means to be human: how they envision their own place and each other's in the hierarchy of humanity, and how they see intelligent machines influencing this. This difference stems from the fact that these two themselves occupy very different rungs on the ladder. As Gurl puts it: 'You know, just cos we've both got pussies don't mean you and me got anything in common. Schooling and money gives you the edge. Makes you more of an important human than me.'[27] Part of what is afforded by Scarlett's privilege is the power to speak with a claim to objective, universal validity. From this standpoint, sexbots are bad because they perpetuate behaviours that have no place in an ideal world of liberty and equality. But Gurl's world is not ideal, and her subjective view, which Scarlett deprecates and denies, is that the robot makes her life better. Scarlett might be right to highlight the problems of oppression and alienation, but the only solution to that problem is for humans to grant each other the humanity they deserve – something she fails to do for Gurl. She grasps the ultimate problem, but not the ultimate solution: to give all humans the kind of respect and friendship that Gurl currently receives only from Roxanne.

3

while (alive) {love me;}

Kate Devlin

The first thing to know about sex robots is that there are no sex robots. Despite Pygmalion's prayers, despite centuries of longing for the perfect created partner and countless science fiction books and films – there are no sex robots. It's not for lack of trying; right now, as I write, there are workshops in the United States and workshops in China, where they are affixing animatronic heads to the stationary and immobile bodies of life-sized dolls, but there are still no sex robots.

It is hard to define a robot by appearance. Look around at those we do have: mechanical limbs on factory floors, programmed to twist and arc; millions of little discs of plastic and metal scurrying across carpets, sucking up dirt; proxy tools for surgeons that slice precisely into skin; heavy, trundling blocks on caterpillar tracks rolling towards a suspect device. Some robots among us may be given humanlike design but, in reality, they do not look anything like us. The robots among us are not designed to invite desire.

The headlines would have you think otherwise. The media loves a salacious narrative and so we see dystopian forecasts of, for example, fembots poised to take over, to fulfil men's every sexual need and put an end to our human–human relationships.[1] If a programmable female machine can satisfy a man's every whim then what is the future for women? These concerns hint at real social issues: an epidemic of loneliness, a longing for the ideal companion, and the fear that women pose a threat to a patriarchal society.

The closest we have today to sex robots are expensive, humanlike dolls with some degree of mechanisation and artificial intelligence (AI) chatbot capability,[2] but the discussions around their potential existence reveals much about ourselves as humans. The debate around the possibility of intimate relationships with machines is important and timely. Social concerns about objectification, legality and privacy reach beyond the topic of relationships with robots. They apply to all our interactions with technology, from social media to video games, voice assistants, and smart devices. They ripple through our interactions with each other, forcing us to question how we relate to living, breathing humans.

Our world is designed for people. We have built environments suitable for ourselves as the dominant species: human-sized doors and paths, stairs the right size for human legs, shelves within reach of our arms, and tools that fit our hands. Ideally, it would make sense to have robots that could fit seamlessly into our environment. Alas, it is very, very difficult to create a human-shaped robot.

Since ancient times the artificial human has long been a compelling tale,[3] and no wonder, since it's easy for us to imagine mechanical versions of ourselves doing the jobs we don't want to do. In practice, however, the engineering challenges are enormous. Since the 1970s, there have been a number of humanoid robots developed both commercially and for research purposes.[4] Some, like Hiroshi Ishiguro's Geminoid series, are intended to be ultra-realistic; others, like Aldebaran's

NAO, are small, open source and cheap enough to be used widely for research and educational purposes. However, it is incredibly difficult and costly to get robots to do physical things we take for granted, such as walk, balance, pick up things, or touch with varying degrees of pressure. It is computationally expensive and – more importantly for developers – it is financially expensive, too.

Even if we *could* build efficient human-shaped robots, could we make them convincing enough? In 1970, robotics professor Masahiro Mori proposed the concept of the 'uncanny valley' (*Bukimi no Tani Genshō*). His hypothesis was that we empathise with machines that have human attributes, but only up to the point where they are close to indistinguishable from reality. At that point, the 'almost-human' robot becomes unsettling and uncanny and fills us with revulsion.[5] The uncanny valley effect is subjective – empirical evidence is inconsistent – but is widely reported.

There are a number of theories as to why we experience this feeling of the uncanny. A significant factor may be that 'human-looking but not alive' is redolent of death. Mori compared it to viewing a corpse: not a pleasant experience, and one that could be frightening, and if it were to move, we would be terrified. But it's not just the realistic-but-not-quite-human appearance of a robot that can disturb us. Coupled with not-quite-human movement, an eerie new dimension emerges: the uncanny valley could be an unsettling reminder of our own mortality.[6]

Conflicting perceptual cues cause cognitive discomfort. Current humanoid robots often have mismatched actions: speech that does not sync with mouth movement, the absence of subtle affective cues, or facial expressions that convey a mood that does not match the conversational tone.[7] We remain stranded in the uncanny valley – the few humanlike robots that exist today could never be mistaken for a real person and it may be that they never will. Tinwell et al.'s 'uncanny wall' theory[8] suggests that, as technology develops, we will become more and more attuned to the subtle mismatches between human and human-created. We might never bridge the divide. It may be easier, someday, as technology advances but, for now, it remains a difficult, if not untenable, task.

Why, then, if the task is so far from possible, are we attempting to build artificial lovers? Studies on companionship and rapport show that it takes very little interaction to elicit a positive human reaction to robots and AI.[9] As social animals, we respond quickly to any technology that engages us. Even an exchange with a disembodied piece of software such as a chatbot is enough to form a social bond if that software is kind to us. (By contrast, we are infuriated by machines that chastise us, which reveals just how social we are.) People are willing to form social and emotional relationships with a robot, even when they know it is nothing more than a machine – even more so if it has humanoid features.[10] It makes sense to us to interact with responsive machines as if they are capable of understanding us, even when we know they can't.

Rapport, then, seems to be key. When that rapport is bundled into a familiar form, and that form is made desirable, then we have our artificial lover. And we find, incredibly, despite being twenty years into the twenty-first century, that form is a reductive and clichéd stereotype of a woman. But then, should we wearily expect anything else when we have been fed endless films of seductive fembots,[11] and when Silicon Valley remains dominated by straight white men who assume the user is just like them?

The prototypes of sex robots emerge from the lineage of the sex doll and prompt accusations of a female objectification that is already rife: women face body-shaming and criticism every day in media, advertising, film, and music. It is no surprise that we tend to gender humanlike robots along socially convenient binary lines.[12] Since manufacturers are almost always creating sexual companion robots for heterosexual men, it is somewhat inevitable that these machines are designed around what already sells in doll form.

Yet despite the highly sexualised design of these would-be robots, manufacturers are selling companionship. Their websites shy away from the erotic and instead promote the romantic:[13] their 'robots' and the conversational AI centre on relationships. 'Your perfect companion' is there in the very title of their home page. Customers are encouraged to 'stay intimate; fall in love'. Artificial companions are to be cherished rather than ravaged. Indeed, not all owners have sex with their dolls, and not all emulate relationships with them.[14] For those that do, the majority speak of their synthetic companions with warmth and respect.[15]

This is where the headlines fail. Sex sells newspapers but is secondary to companionship when it comes to artificial lovers. Lurid tabloid ledes that declare sex robots a threat to humans are playing into much broader and long-standing fears around automation and loss of agency. How could we not be frightened that robots might not only replace us in our jobs but also in our beds? And the narrative of the murderous fembot? We can perhaps interpret that as the cautionary tale of women 'breaking their programming'; finally fighting back against the status quo.[16]

The future of human–machine relationships does not have to centre on a robot shaped like a life-size Barbie doll.

Manufacturers say that male versions of these prototype sex robots exist,[17] although this sounds tokenistic at best. If we can form social bonds easily with technology, do we really need the machine to look human? Does it even require a body? The rise of more sophisticated chatbots has led to the development of virtual avatars – no physical body with its cumbersome mechanics, instead, an artistic or realistic rendering of a human form with a conversational AI. There's still no escaping the gendering,[18] but, for many users, gendering may play a significant role in attachment. An added benefit is that in virtual form, it is much easier to escape technological limitations: thus the uncanny gap is narrowed.

The virtual companion is a much more readily accepted form of artificial partner since voice assistants already carry out commands and the taboo is lessened. (On a practical level, a virtual companion is also a lot cheaper and much easier to store.) We have already accepted conversational AI as a useful tool – we speak to chatbots all the time. Sometimes we don't even know that we are speaking to them. There are anecdotal reports of emotional relationships, too: a sense of companionship with the AI assistant – 'not a fake, but a phantom friend.'[19] Gatebox, a Japanese company who make a voice-driven AI with an associated projected holographic anime character, say they have issued thousands of certificates for men who have 'married' their virtual assistant.[20]

Developments in affective computing allow AI and robots to read our bodies, scan our faces, and interpret our moods.[21] When the computer knows how we're feeling, it responds accordingly. We don't have to feel tricked or duped: we can accept them as machines, yet still feel understood, even when we know they aren't conscious or sentient. Therapy bots and companion pet robots are testament to this empathic effect.[22] For those who wonder if we can really love something that doesn't know we exist: there are millions of people out there every day, gazing on humans they love from afar, who are unaware of their feelings.

Alongside the virtual companion, sex tech brings us a dimension of pleasure. Sex toy development in the past few years has diversified into forms and shapes suited to all types of bodies.[23] Some of this is customisable and personalisable, and smart-connected devices can be controlled via Bluetooth or the Internet by a partner(s). Sexual and sensual experiences can be created via augmented and virtual reality. It is a multi-billion-dollar market and is becoming more widely accepted and better funded every year,[24] and the potential for imaginative development is profound.[25]

Concerns that rapport with machines will isolate us have endured for centuries, each time new technology is introduced: writing, the printing press, newspapers, television, smartphones.[26] We humans are amazing at adaptation. We are already adapting to a world with AI and we have hardly noticed it.

A human–machine relationship does not have to resemble a human–human relationship. A physical robot or a virtual AI, does not have to be a replacement for a human. It *could* be, but it could also be something else, something new in its own right. For those who say it's wrong to want to be close to a piece of technology, I would ask why it's so important to be close to another human? Who are we to judge how others lead their lives? If someone finds solace, pleasure, or companionship in a non-conventional way, and causes no harm to anyone, what's the basis for objection? For some, it may be that a human–machine relationship is sufficient, or even superior. For others, it may be a proxy if human companionship is not available. And, for many, it could become another friend, companion, or partner in an extended friendship group of humans (and maybe other humans' machines) – relationships that could enhance their lives.

Predicting the future is a fool's game. Each time someone working in AI talks of 'twenty years from now', it's a given that in twenty years' time they'll be asking for another twenty years. The future is always just over the horizon. But history shows that the dream of the artificial companion never disappears because it is a story of being loved unconditionally, being understood by someone belonging entirely to you. It is a human story but it is only now that we see it being realised, as technology is delivering the beginnings of a tangible experience. It is not yet ready, it is not yet perfect – and there are still no sex robots. But there are watches that read our heartbeats and cameras that capture our expressions, and millions of people thanking their robot vacuum cleaners and laughing at their voice assistants and using their phones to call their families and tell them they miss them and love them. If we can mediate our relationships through technology, then it is not too far-fetched to think about a day when we can be in relationships *with* our technology.

4

Robots as Solace and the Valence of Loneliness

Julie Carpenter

Robots are notoriously challenging to explain in one universal definition. However, they can be defined collectively, in part, as intelligent machines that can learn from and interact with their environment with some level of autonomy. Sub-categories include robots whose functionality is intended for industry, service, home consumers, the military, or entertainment purposes. Each category indicates completely different design intents and anticipated use scenarios; thus developers conceptualise each robot's purpose and design accordingly.

Robots are becoming increasingly common in the everyday world outside labs and industrial settings. Social media entertains us with clips of dancing robots, doglike robots slipping across ice and nimbly walking over rocks, healthcare robots for elderly people, even humanlike robots used in space exploration. We see consumer drones flitting about in neighbourhood parks, floor-cleaning robots in our homes, and in some cities, delivery robots zip along sidewalks or greet us in shopping malls. In other words, robots take many forms with different primary functions and how we view them in relation to us is always changing. Culturally, all around the world we are navigating how we should interact with these intelligent machines whose reputation precedes them in our real world.

Robots are sometimes designed with recognisably human or animal-like social cues in order to leverage our mental model of how to interact with them. These cues can include aesthetics such as the shape and anatomy of the robot (such as if it has 'eyes' or limbs), the situation in which you interact with the robot and its role in relation to you, and its ability to communicate with language or voice. For example, talking to a robot builds on our everyday model of how we communicate with people. Yet leveraging our mental model of human–human (or sometimes human–animal) interaction and moulding it to fit human–robot interaction also creates a set of interesting and evolving cultural phenomena. Robots are relatively new objects in our world, but they have become a strong cultural touchstone for many people because of their mythos as characters in popular storytelling.

In science fiction, when a character interacts with a robot as if it was alive and social, we embrace the fictional relationship. Indeed, the core of the human–robot story arc depends upon recognisable social interactions. In *The Wizard of Oz* (1939), the Tin Man becomes a beloved friend of the protagonist Dorothy; in *Star Wars* (1977), the robots C3PO and R2D2 have become folk heroes to generations of moviegoers. Even dystopic human–robot relationship portrayals, such as in the *Terminator* film franchise, depend upon the audience understanding why the robot and human characters interact the way they do.

How people interact with robots 'in the wild' will be influenced, in part, by their own fluid ideas about how robots should be treated, interacted with, and regarded. We often treat them as tools, interacting with them as objects used to accomplish a goal or task. But people also sometimes view robots as social Others, whether or not they were designed for that purpose.[1] Much has been written about the multiple factors that appear to influence how we treat a robot socially: morphology, movement, level of intelligence, language use/understanding, and the situation in which the human–robot interaction is taking place. By contrast, little has been written about changing cultural attitudes towards robots and how these influence human–machine relationships.

In particular, there has been relatively little scholarly exploration of platonic emotional intimacy between human and robots. Yet the burgeoning body of work that demonstrates human–robot attachment possibilities indicates that this condition should not be dismissed as niche or, indeed, as deviant human behaviour that results from loneliness or psychological vulnerability. My own research suggests that attachment may be a very normal human reaction to a robot in certain situations,[2] triggered by a combination of influences including the cultural expectations and beliefs of the user, their cognitive model of human-to-robot interaction, design cues from the robot that afford (previously exclusive) organic or humanlike modes of interaction (e.g. natural language processing), to the persistence of the human–robot relationship over time.

I spent several years researching explosive ordnance disposal (EOD) personnel in the US military to study how they interact with the bomb-disposal robots they use as critical everyday work tools. In addition to the unique physical risks of their profession, like other military personnel, ordnance technicians are often placed in stressful situations that can leave them emotionally isolated from family and without established emotional support systems for extended periods of time. Although often surrounded by friends, co-workers, and peers, loneliness can be the result of not just a lack of *enough* interpersonal relationships, but a deprivation of emotionally intimate ones.

Analysis of my research with these soldiers revealed several behavioural patterns, including that they had frequently created social categories for some of the robots they used in their work. It is important to understand these robot models appear very machinelike – instead of wheels or legs their mobility is tracked, so they look like small metal tanks and do not in any way resemble humans or animals. One model in use at the time of my research was approximately the size of a backpack; another was similarly shaped but larger and meant for different situations. Neither model was weaponised, as their use is to assist soldiers in making safe unexploded ordnance at a distance. Neither of these robots were responsive to human language or gave lifelike social cues via their design. They were semiautonomous (i.e. operated remotely by a soldier), designed to function as tools and not designed for any social purposes. Additionally, EOD are, of course, highly trained and skilled personnel whose work requires rigorous attention to detail, focus, interpersonal communication, and technical skills.

Yet, of the soldiers I spoke with, some identified meaningful emotional attachment toward robots they had worked with for a period of time. A specific and significant pattern also emerged of their assigning robots personas. One soldier described the naming ritual for the team, 'We'd name them after movie stars we'd see in the theatre, or music artists, someone popular, and then we'd always go to vote to decide on'.[3] Others described naming the robot after their real-world partner, and associating human personality traits with the robot. One soldier commented, 'I've spent more time with the robot than my girlfriend in the last six months', and described how they slept next to the robot in the truck 'like a girlfriend'.[4] These are very rudimentary associations – and by that I mean the soldiers did not truly see the robot as a *substitute* for a meaningful human relationship. However, they did interact with their machinelike robot as a social Other, and they did, as a group, agree on *acceptable* ways of interacting with it socially. This determination of and consensus about *appropriateness* is dynamic and not necessarily formally discussed, as with the team voting for the robot's name. This new etiquette constitutes a set of behaviours the soldiers have tacitly decided upon and agreed among themselves as acceptable when interacting with this technology.

After my work about the soldiers was initially published, other military personnel reached out to me to share their own experiences with robots. I received the following email from a retired bomb technician who shared this very personal story:

> I recently read a small article about your research into EOD personnel and their attachments to their robotic platforms. As I am an EOD technician of eight years and three deployments, I can tell you that I found your research extremely interesting. I can completely agree with the other techs that you interviewed in saying that the robots are tools and as such I will send them into any situation regardless of the possible danger.

> However, during a mission in Iraq in 2006, I lost a robot that I had named 'Stacy 4' (after my wife who is an EOD

tech as well). She was [an] excellent robot that never gave me any issues, always performing flawlessly. Stacy 4 was completely destroyed and I was only able to recover very small pieces of the chassis. Immediately following the blast that destroyed Stacy 4, I can still remember the feeling of anger, and lots of it. 'My beautiful robot was killed ...' was actually the statement I made to my team leader. After the mission was complete and I had recovered as much of the robot as I could, I cried at the loss of her. I felt as if I had lost a dear family member. I called my wife that night and told her about it too. I know it sounds dumb but I still hate thinking about it. I know that the robots we use are just machines and I would make the same decisions again, even knowing the outcome.

I value human life. I value the relationships I have with real people. But I can tell you that I sure do miss Stacy 4, she was a good robot.[5]

This story is an interesting example of how a soldier projected an emotionally rich social narrative on to a machinelike robot not designed for social interaction. His experience elucidates his thought processes, which include his self-awareness of a poignant mix of sense of loss of the robot and embarrassment that he had an unexpected emotional connection for the 'tool'. At the same time, he explains the robot is, to him, its own social category and that he would not hesitate to make decisions that endanger the robot. In his eyes, the robot is absolutely not valued similarly to *human* life; it is in its own social category. The robot became a vessel for him to express versions of solace, comfort, safety, humour, and even loss, which were meaningful for him in those situations but not necessarily a substitute for human forms of socialness.

Culture is dynamic. There are currently very few social rules widely accepted about our etiquette with robots as social beings. Formal social structures, such as religious and legal doctrines concerning human–robot relationships, have not been consistently or broadly established. There are discussions and questions about whether it is an *immoral* act to feel friendship or socialness toward a robot and whether people are substituting technology for human relationships to their own detriment.[6] However, as with expressions of physical affection, a discussion of emotional affection – and possibly, attachment – centres on what are considered *appropriate* behaviours. The idea of socially acceptable human attachment to a robot is less easily understood or even researched because internal motivations, such as intent, are not necessarily clear, even to the principal social actors in the scenario.

History is littered with examples of censure and hostility towards people who have engaged in relationships that were deemed unusual at the time. There is a clear tension between the desire to create a connection or sense of belonging and the knowledge that your chosen relationship may be critically judged by others. We have arrived at a cultural stage with human–machine relationships that centres on this tension, in many ways. To admit that a robot has become socially meaningful to us is unlikely to meet with positive acceptance – and very likely to prompt derision and hostility. Acknowledging that a robot can provide a positive and comforting emotional connection is a difficult thing for many people to talk about openly or casually. It is still common to make jokes about people who reveal their affection for a robot, or to dismiss such relationships as an invalid social outlet or even a dangerous or deviant inclination.[7] But as robots increasingly become part of our everyday lives, our feelings about human–machine relationships will become more nuanced. In some cases, we will regard robots as their own set of emerging social categories.[8]

Loneliness is a common feeling for many people, whether it is temporary experience or an ongoing sense of being alone. It is acknowledged as a distressing experience with potentially serious emotional, mental, and physical consequences.[9] Loneliness is not necessarily caused by a deficit of desired interpersonal relationships – it may, indeed, prompt a person to seek social interaction with a robot. The tension between the human need for social connection and the knowledge that the robot is a machine and a medium for their own projected social needs can position it as a potentially tempting blank slate for a creative – and perhaps in some cases even therapeutic – outlet. People in emotionally vulnerable situations associated with loneliness may also be more open to the idea of viewing a robot in a social way. Social relationships are at the core of human life. Historically, there has been a commonly held assumption that people who believe that they enjoy a meaningful relationship with a robot may be socially maladjusted or exceptionally lonely. However, a shift in our cultural belief systems and attitudes is underway, prompted by an increased exposure to robots that is nuanced, subjective, and situation-dependent. In fact, our desire – or instinct – to interact with some robots socially may be one of the most human things we naturally do.

5

Robot Nannies Will Not Love

Joanna J. Bryson and Ronny Bogani

Childcare is the most intimate of activities. Evolution has generated drives so powerful that we will risk our lives to protect not only our own children, but quite often any child, and even the young of other species. This compulsion is evident across the animal kingdom where we see, for example, birds feigning wing injuries to distract predators from vulnerable chicks.[1]

Robots, in contrast, are products created by commercial entities with commercial goals which may – and indeed should – include the well-being of their customers, but will never be limited to that.[2] The protective instinct that overwhelms a parent (or any person) when they see a child in danger is evolution focusing our attention in a way that is historically proven to best propagate a version of our society and our species into the future.[3] Robots, corporations, and other legal or non-legal entities do not possess the instinctual nature of humans to care for the young. A robotic caregiver such as a robot nanny may be programmed to mimic or simulate human expressions of empathic behaviour but it does not *feel*: it cannot experience emotion or desire, at least not in a way remotely like an animal would.[4] However, in the way that we attribute emotions to our pets, our anthropomorphic tendencies may prompt some children and adults to overlook this fact.

This chapter addresses the legal framework in which commercial childcare robots will be marketed and our concerns about the likelihood of deception – both commercial deception through advertising and also self-deception on the part of parents. We also examine the social impact of robot childcare and explain why robots are unlikely to cause significant psychological damage to children and to others who may come to love robots.

We must first dismiss out of hand the idea that robots themselves have any liability. Robots are tools purchased and owned by parents (or other human legal entities such as the state or commercial nurseries) and marketed by corporations. Any effort to introduce a new layer of responsibility at the level of product just because a robot nanny has humanlike features should be considered a de facto evasion of responsibility, most likely on the part of the manufacturer.[5]

Even though we must know that a robot cannot feel like a parent feels, and will always act at best as a surrogate – not only of the customer who bought it, but of the corporation that built it – we can nevertheless expect that a robot purchased for use as a nanny is unlikely to ever do much damage to children. Except, of course, in a few unfortunate cases – just as children are, sadly, hurt or injured by human nannies – and for that matter, parents.[6]

Robot childcare will require codes of practice and possibly legislation to enforce realistic advertising and warn about appropriate use. However, we predict that this legislation could very likely come about as a result of manufacturers *under*-selling the artificial intelligence (AI) and interactive capacities of their robots. No company will want to be liable for damage to children. It is therefore likely that any robot (and certainly those in jurisdictions with strong consumer protection) will be marketed as a toy or monitoring device.

Neither television manufacturers, broadcasters, nor online game manufacturers are deemed liable when children are left for too long in front of their television. Robotics companies will want to be in the same position, so we should be able to communicate this model to them: make your robots reliable, describe what they do accurately, and provide sufficient notice of reasonably foreseeable danger from misuse. Then, apart from the exceptional situation of errors in design or manufacture, such as parts that come off and choke children (a common risk among such device makers), legal liability will rest entirely with the parent or responsible adult; as it does now, and as it should under existing product liability law.[7]

Consequently, we can see that the primary responsibility of robotics professionals and experts – such as ourselves – will be educating parents on the proper use of robots, and doing research on their impacts on children. As with any other new child-oriented strategy or technology, we can expect that information on robot nannies will rapidly disseminate through chat shows, social media, online parent chatrooms, and magazines. There will be a plethora of theories, articles, and received wisdom about the right and wrong way to use robots with children. We can expect widespread controversy and debate. Perhaps this will even result in a cottage industry of professionals dedicated to safely maximising contextually specific functioning of domestic social robots.

What robotics manufacturers will need to worry about is that robots will be banned due to incredibly rare cases of neglect or misuse, as with extremely popular children's toys such as assisted walkers and lawn darts. The failure of a small number of parents to adequately safeguard stairways or properly supervise their children has resulted in injuries and/or death.[8] However, rather than the parents being blamed, these products were quickly banned. No legislator wants to be associated with dead babies. But legislative banning is not inevitable. Not all products are created equal and certain products, such as guns or automobiles, are dangerous instrumentalities by design. But no one has yet banned these far more significant causes of child injury and death.[9] Automobiles are seen as too critical to our economy and our individual freedom to be banned, despite the horrific cost in loss of life resulting from negligent use or design. Guns are (in some countries) afforded political protection for fulfilling their primary function – dispensing death.[10] These kinds of immunities to arbitrary legislative bans could very likely extend to robots as they become more essential economically, politically, and more embedded in our daily lives.

Of course, as more injuries occur through common misuse, the robotics sector will experience regulation by litigation. Enough commonly occurring injuries or legal claims associated with a product will provide notice or warning to manufacturers and regulators of foreseeable risk.[11] For example, the Ford Pinto's rear-fuel chamber design resulted in horrific injuries and death.[12] Civilian lawsuits resulted in punitive damages for injuries, thus highlighting potential dangers and the need for regulatory oversight.[13] Such a litigation model allows our children to enjoy playgrounds, despite the fact that these too are potentially fatal.[14] Notwithstanding the lack of significant economic or political protections, playgrounds continue to remain open despite a multitude of yearly injuries. As long as the majority of robotics companies, their shareholders, and boards of directors remain within the legal jurisdiction of countries with a strong rule of law and a regard for citizen well-being, this model of regulation by litigation will endure.

We predict that childcare robots will be marketed primarily as toys, surveillance devices, and possibly as household utilities. They will be brightly coloured and deliberately designed to appeal to parents and children. We expect a variety of products, some with advanced capabilities and some with humanoid features. Parents will quickly discover a robot's ability to engage and distract their child. Robotics companies will program experiences geared to parents and children, just as television broadcasters do. But robots will always have disclaimers such as 'this device is not a toy and should only be used with adult supervision', or 'this device is provided for entertainment only. It should not be considered educational'.

Nevertheless, parents will notice that they can leave their children alone with robots, just as they can leave them to watch television or play with other children. Humans are phenomenal learners and very good at detecting regularities and exploiting affordances. Parents will quickly notice the educational benefits of robot nannies that have advanced AI and communication skills. Occasional horror stories such as the robot nanny and toddler tragedy in the novel *Scarlett and Gurl*[15] will make headline news and remind parents of how to use robots responsibly, just as they do now with respect to guns, dogs, and wading pools. This will likely continue until or unless the incidence of injuries necessitates redesign, a revision of consumer safety standards, statutory notice requirements, and/or risk-based uninsurability, all of which will further refine the industry. Meanwhile the media will also seize on stories of robots saving children in unexpected ways, as it does now when children (or adults) are saved by other young children and dogs. This should not make people think that they should leave children alone with robots, but given the propensity already for us to anthropomorphise robots it may make parents feel that little bit more comfortable – until the next horror story makes headlines.

Introducing a robot nanny into in the family home raises the question of the effect of anthropomorphisation. Manufacturers may seek to maximise or minimise the human features of their robot or AI product.[16] One concern is that excessive focus on exactly the right issues – liability and attachment – may paint such an extreme story of psychological damage that important parts of these warnings may be ignored. Most parents will view robots as similar to television or smartphones, ignoring published medical guidelines; but they should not assume that the robot itself will provide sufficient care. Problems of neglect associated with robotic care and parental absenteeism will likely be largely the same as they are now with televisions – at least qualitatively, though AI may allow for quantitative changes in prevalence of harms. Robots will in the majority of cases be used to complement human care than replace it. Adults will always be required to handle novel or emergent situations and determine each child's individual capability to interact with robots and the environment in a safe and appropriate manner. There will be parents who will use and perhaps overuse robots as child minders, but this is not significantly different from the current situation with smartphones and other AI devices. What may change is how much attention that responsible person can devote to other things, or how many children that person can manage without stress or exhaustion.

Robotic childcare will have significant implications for childcare workers and wages. Nurseries may be able to support more children, and more people may be attracted to the field or prove competent in it with the addition of AI assistance. This could reduce earnings in what is already a low-paid occupation, or of course it could just improve the quality and therefore the perceived value of the service. Eventually most societies should come to value important labour even if it was previously provided for free primarily by women in domestic households, but this is a wider problem for many service industries. We would hope that parents and taxpayers would value childcare and other AI augmented human services and believe that good wages and investment in good technology are essential. Society needs to become more attentive to technology's potential for improving the human condition rather than simply focusing on more immediate payoffs like wealth and immediate consumption.

Robots are very different from television and dogs in that they provide interactivity of a highly reliable sort. While this extreme reliability can be partially ameliorated by artificial emotions and noisy sensing,[17] ultimately children will realise that robots are more predictable than humans. Robot nannies will not be irritable, they will not lose their temper or forget or ignore, and they are available 24/7. Just as television and social media increase the probability that children will have attention deficits,[18] robots may increase the probability that children develop bonding issues with their parents and friends. Some children will prefer the more reliable style of interactions they find with machines – just as some prefer simpler interactions with animals or the high-bandwidth, low-risk stimulation of books.

We know that a child's ability to bond with parents has a long-term impact on their ability to form friendships, romantic relationships, and generally integrate with society.[19] Our concern is that children who prefer the company of predictable interactions with robots may be setting themselves up for a life-long preference for machines over humans. This important possibility can already be explored experimentally, by looking at children and adults who prefer AI versus human opponents in online gaming. Yet, to our knowledge, this research has yet to be done.[20] Nevertheless, another legislative direction that might plausibly emerge with respect to robot nannies would be mandatory warnings about how addictive these technologies can be for some children, or recommendations about time limitations for robotic exposure and engagement. However, despite many academic studies recommending limits in exposure to television and computer games, no such legislation has yet been written for television.[21]

Of course, even if we discover correlations between children with a preference for interaction with AI and robots and the expression of other forms of introverted behaviour, this does not mean that AI and robots are necessarily bad for these children. Indeed, these devices may provide stability and comfort that reinforce some children's sense of self-worth. Given how little is known about children and AI, we should not overlook the possibility that robot nannies and AI toys might be beneficial in unexpected ways. There is, for example, the chance that protracted experience of AI might enhance a child's understanding of themselves and what it means to be human.[22]

The United Nations Convention on Rights of the Child, which has been ratified by every nation in the world – with the notable exception of the United States – enumerates certain basic human rights for each child.[23] A primary right of this all-encompassing social, political, economic, and cultural treaty is the concept of a child's agency and right to participation.[24] This right to be heard ensures children's right to have their opinion considered in matters that affect them.[25] Gauging the individual capability of each child to engage in certain activities may be augmented as it is now by technology, but will remain a parental duty. Real-time ongoing evaluation of a child is not delegable to a robot or the corporation that manufactured it. Nevertheless, information derived from interaction with the robot may inform both the child or the parent. And, of course, that information may also be used to inform *on* both child and parent.

Robotics and AI have already entered the domestic domain in the form of smartphones and AI assistants, and research reveals their impact on traditional family dynamics.[26] For example, it will be more difficult to protect or restrict children from perceived danger if a child can use their AI to present empirical data to the contrary. As children mature and grow in their understanding of AI and robotics, we can expect they may convert these devices into personalised legal and advocacy tools. This could empower a child or young person to assert and defend their rights, and enable them to challenge and override parent/guardian restrictions. Such a development would depend on legal protections concerning the transparency and accountability of AI products being equal for products through all stages of childhood.

Of course, this is not to suggest children will necessarily be wiser than their parents. They will always be at least as susceptible to advertisers and political manipulation. Nevertheless, our goal must be that AI empowers not only the adults who love children, but the children themselves. While this may challenge traditional parenting techniques, it will likely not harm the child or young adult. Rather, or perhaps also, AI may help build resilience and prepare them for the challenges of the human condition.

While we have argued somewhat dryly here about a matter of passionate importance, we have made a number of predictions. We believe that manufacturers will under-represent the capabilities of childcare robots in an effort to avoid liability. Children will be changed by interacting with robots, and indeed probably already are being changed by AI. But we believe it is unlikely that robots will cause signifi-cant psychological damage. As academics we have ethical obligations to prepare society for the changes that research elicits, though sadly, even leading experts can be sources of deception (including self-deception) about the capacities and desired cultural roles of robots. We need to accelerate the rate of public engagement in and understanding of AI and robotics. As individuals, we need to re-examine our goals and the out-come of our work, purchases, and advocacy on our society, and particularly, our children.

The bottom line though is that the robots themselves will not love our children. Robots are manufactured devices that if they represent anything, represent the entities that develop and market them. With adequate regulatory concern they should also represent the needs and desires of parents and children as well. And as with many far less engaging toys, we must always remember, the children will love their robots.

6

Masters and Servants: The Need for Humanities in an AI-Dominated Future

Roberto Trotta

20 February 1947: It is a bitterly cold day. As he sets off from his home in Hampstead, North London, the man clasps the collar of his overcoat against the wind, his boyish looks disguised by his woolly scarf and hat. Ice fields have been reported off the Suffolk coast, and the forecast is for more snow and freezing temperatures across the country. He hopes that the further blackouts expected today won't disrupt his lecture. He needn't have worried: over a hundred scholars – double the usual number – will attend this monthly meeting of the London Mathematical Society, no doubt seduced by the beguiling title of his talk: 'The Automatic Computing Engine.'

Daylight is fading already in the grand Royal Academy courtyard as the man pushes open the heavy wooden doors of Burlington House. His hand (we imagine) clutches a brief-case with the thirty-one typed sheets of his lecture and he runs, once again, through the central plank of his argument: that it ought to be possible to build a machine that won't be bound by a fixed set of commands. Instead, the machine should be able, 'if good reason arose', to modify its instructions to achieve its goals more efficiently, in a way that might eventually escape the understanding of its human creators. At this point, one would be obliged to regard such a machine that can learn from experience as 'showing intelligence'.

13 March 2016: The Grandmaster hesitates in the windowless, air-conditioned room, a white stone delicately held between his fingers. He looks younger than his age and an aura of intense concentration radiates from his eyes as he studies the almost infinite possibilities of the gridded board. He rises suddenly and steps out on the empty terrace overlooking Seoul, menacing clouds gathering over the city. He smokes nervously, plotting his next move.

Oh, how confident he had felt going into the series of games! Foolishly, he even went on record saying he expected to win 5–0. Now, three games in without a single victory to his name, the Grandmaster badly needs a break – not just for himself, but for humanity's reputation.

These two men never met: Alan Turing's life was cut dramatically short in tragic circumstances that have never been fully explained, almost three decades before Grandmaster Lee Sedol was born. Turing fired the gun of the artificial intelligence (AI) race with his visionary ideas about thinking machines that could surpass their masters; Lee, for all of his eighteen international titles as a Go grandmaster, was defeated by the devastating moves of a machine that wrongfooted a human intuition shaped over thousands of years of practice of this ancient Chinese board game. In just under seventy years, Turing's thinking machines had evolved from a slumbering pachyderm of valves, fed with punched cards by an army of human 'servants', with memory banks (so he had envisaged) in the form of sound waves travelling in mercury tanks, to the inscrutable workings of reinforcement learning taking place in the cloud. When Lee Sedol was pitted against DeepMind's AlphaGo code in a series of five games, with a million-dollar prize at the end, all that he could do was to bring home a single victory for humankind, thanks to a brilliant move in game four that for a moment tangled up the carefully optimised weights of AlphaGo's deep neural network.

The machine had been, perhaps for the last time, beaten by its master.

No doubt Alan Turing would have been pleased to see a descendant of the machines he dreamt about excel at the game of Go. He concluded his 1947 lecture to the London Mathematical Society by proposing chess as a good testing ground for his machines. At some point, Turing must have begun to think about Go as an even more daunting challenge, since two undated cream-coloured sheets of thick King's College paper found after his death bear, in the mathematician's neat handwriting, the rules of Go summarised in seven numbered points.[1]

Three years after his lecture, the father of AI envisaged building child machines (i.e. machines with a malleable brain) that could be taught by a process of 'punishments and rewards', foreshadowing what is today known as reinforcement learning – the very technique at the heart of AlphaGo's victory.[2]

Three years after his 4–1 defeat against the machine that Turing had dreamt of, Lee Sedol retired from professional Go playing, having lost his drive to excel in the game now that a non-human 'entity that cannot be defeated' was towering over him.[3]

The machine had fulfilled the high expectations of its father.

Today, Turing's thinking machines are helping astrophysicists such as myself advance our knowledge of the universe. But AI has spread and thrived outside academia, and, while often difficult to spot, it is now practically everywhere: it chooses the next movie or song we will stream, decides what we see on social media, seduces us with purchases that fulfil needs and desires we didn't know we had; it eavesdrops on our intimate conversations, which are beamed back to the base and shared with human contractors without our consent;[4] AI-powered virtual assistants are eager to please us by telling us the time, reciting our daily schedule, or ordering us pepperoni pizza, all the while reinforcing gendered stereotypes and racial biases with their default feminine, white, educated voice.[5] Turing's prophecy that the 'servants' of his machine, women (one has to assume) that 'feed it with cards as it calls for them', would eventually be replaced by 'mechanical and electrical limbs and sense organs' has come true, if at a high environmental and societal cost. The omnipresence of AI is supported by an invisible infrastructure of power-hungry data centres and manufactured by an exploited workforce in dismal electronics factories.[6]

In the near future, AI promises to drive our cars, reduce our electricity consumption by cracking difficult optimisation problems, diagnose illnesses earlier and more accurately than human experts ever could, and discover new drugs at a vastly accelerated rate. Sooner than you might think, AI may also autonomously assassinate targets from the sky, undermine democracy by stealth disinformation, erode public trust by creating fake videos indistinguishable from reality, and underpin mass surveillance methods that will make authoritarian rulers unassailable.[7]

Just as the taming of the atom gave us both clean energy and immense destructive power, Turing's thinking machines have the potential to help humankind develop a more equitable and prosperous society, or to amplify current disparities in wealth and power. This is not lost on governments, academics, and some industry leaders, who in various fora have put forward sets of ethical principles to guide AI development. Common among several of the emerging frameworks are the requirements of beneficence (promoting human well-being and planetary sustainability), non-maleficence (protecting privacy and security), autonomy (the power to take decisions), justice (avoiding unfairness and promoting prosperity), and explicability (intelligibility and accountability).[8] However, two further aspects are crucial to help usher in an AI-dominated future that works for the common good: abandoning the view that AI is solely a computer science problem, and achieving a more diverse representation in the field. Both of these aims require a view of AI as part of an interdisciplinary context with complex ramifications into all aspects of society, not just as a technical challenge. In other words, they require a humanistic approach to science.

The gap between the scientific and humanistic cultures, famously decried by C. P. Snow in his 1959 Rede Lecture,[9] has been growing wider. Paraphrasing Snow's comment about nineteenth-century physics, one could say today that the great edifice of twenty-first-century AI goes up, 'and the majority of the cleverest people in the Western world have about as much insight into it as their Neolithic ancestors would have had.' The reverse is also true: the cleverest (and most successful) people in Silicon Valley are often steeped in a narrow monoculture that sees an app as the solution to every problem, blithely unaware of its wider and subtler cultural, historical, and societal dimensions. The cultural homogeneity and lack of a broader, more inclusive perspective in the Silicon Valley tech scene,[10] means that hugely consequential decisions for a large majority of the world's population are taken by a narrow group of like-minded individuals. Examples of what has been called 'technochauvinism'[11] include racial and gender biases built into training data for AI systems (e.g. for facial recognition[12] and crime prevention[13]), misuse of personal data and violation of privacy (as in the Cambridge Analytica scandal), and AI-empowered systematic discrimination and disenfranchisement.[14]

In my experience, progress can be made by reintegrating the study of humanities into scientific and technical education. As a theoretical cosmologist with expertise in statistical methods, in recent years my research has focused on data science and machine learning; at the same time, I was leading and embedding humanities education in my role as director of the Centre for Languages, Culture and Communication at Imperial College London (2015–20). Like C. P. Snow, it was a 'piece of luck' to find myself with a foot in each camp at a time of such fast-paced change.

I have thus witnessed first-hand the transformative power of the humanities, arts, and social sciences on some of the brightest young minds from all over the world, reading towards science, engineering, and medicine degrees at Imperial, one of the top ten higher-education and research institutions in the world. Already in 1950, the college recognised the importance of 'effective judgement and creative thinking in subjects that lie outside specialised studies and coursework.'[15] This long tradition blossomed over time, and Imperial students today can choose from more than 180 modules in the humanities, languages, and global challenges. These modules are designed to expose students to the wider societal context of their chosen field, to expand their horizons, to work in cross-disciplinary teams on 'wicked problems', to make them comfortable with uncertainty, risk, and ambiguity, to become fluent communicators across disciplines and languages, to be aware of ethical and moral aspects of their work – skills and topics that are rarely addressed in a purely technical education in science and engineering.

Students and alumni in science and engineering (including computer science) overwhelmingly report the beneficial impact that modules such as creative writing, music technology, philosophy and law, politics and languages, global challenges and history have had on their intellectual and personal growth.[16] Blind spots are removed, new vistas opened; a mathematical world that was entirely made of binary right or wrong answers becomes infinitely shaded when illuminated by the searchlight of the humanities: notions of culture and values, morals and choice, history and identity, power and politics, masters and servants, and their infinite interconnections with science and technology, emerge from the fog to become the landscape surrounding the narrow technical path of AI. There is no doubt in my mind: scientists, engineers, and computer scientists equipped with an appreciation for humanities are far better prepared to evade the trap of technochauvinism – and will be better leaders in a changing world that needs to navigate the complex and interconnected challenges of climate change, runaway AI development, and widening inequalities.

We are left wondering what Turing would think of his eponymous twenty-first-century machines. Would he have rejoiced to see humans on the verge of being surpassed 'in all purely intellectual fields', as he'd hoped for in 1947? Would he have sought atonement, like Einstein and Oppenheimer did in vain for atomic energy, by campaigning against the unstoppable force he had unleashed? Would he have joined the singularity cheerleaders[17] to hasten the demise of an outdated, purely biological consciousness? Would he have repeated his argument that to 'enjoy strawberries and cream' is 'idiotic' for a machine, or would he have realised that it is part of what makes us human? Would he have happily stood aside as an advanced AI took his place, or would he have been one of the masters 'unwilling to let their jobs be stolen this way', something he considered 'a real danger'?[18]

We shall never know.

On Friday, 4 June 1954, Alan Turing left his office at the University of Manchester for the bank holiday weekend, a note in evidence on his desk with a to-do list for the following week. Three days later he was found dead in his bed, killed by cyanide poisoning. At the time, it was surmised that Turing's mental state was unstable due to the hormone therapy he was forced to undergo as a convicted homosexual – a punishable crime in 1950s England. The coroner pronounced it suicide, though the evidence is brittle by today's standards.[19] The half-eaten apple on Turing's bedside table was presumed the means by which he took his life – but the apple itself was never tested for the poison. Suicide, accident, or even conspiracy? Certainty is now beyond reach.

The apple of AI knowledge that Turing left behind, which he had barely tasted himself, might yet be the most bountiful gift to humankind – or an inadvertently poisoned fruit.

We shall soon find out.

7

A Feminist Artificial Intelligence?

Mary Flanagan

Advances in machine learning (ML) and artificial intelligence (AI) have implications for the future of work, but the rise in automation has wider implications for culture and knowledge: specifically due to the unconscious biases embedded in computational systems. This chapter examines why the under-representation of women working in technology has critical implications for how we use computation and what problems we can address.

The fascination for artificial life has always been part of our imagination, from ancient golem myths to Mary Shelley's famous monster. 'None but those who have experienced them can conceive of the enticements of science,' Mary Shelley presciently wrote in her novel *Frankenstein; or, the Modern Prometheus* (1818). Even from her nineteenth-century viewpoint, writing as a pregnant eighteen-year-old, Shelley critiqued the Promethean fantasy that is characteristic of the Anthropocene, and demonstrated that the seduction of the possible – and the narcissistic act of humans defying natural limitations by creating life itself – can have terrifying consequences. In her novel, Shelley followed her technology-enabled 'fantasy of the possible' to its dystopian conclusion. After the monster awakens and Victor Frankenstein watches the 'lifeless thing' animate, he admits with dread, 'I had desired it with an ardour that far exceeded moderation; but now that I had finished, the beauty of the dream vanished, and breathless horror and disgust filled my heart.'[1]

We dream big dreams about artificial beings but more often than not end up creating monsters. And never have there been more monsters, changing the very standards and structures of civil society, than now, in the tech industry. Silicon Valley's incredibly pervasive influence, not only on the technical structures but also the societal values of AI, affects how and what a typical global citizen might encounter on a daily basis. From facial recognition-based profiling to internet searches, from social media algorithms to maps, the logics of AI systems influence our experiences of the world.

However, women are not truly participants in the design of the technologies that shape our world. Women still only earn 18 per cent of the undergraduate computer science degrees in the United States, and 13 per cent in the United Kingdom, despite surging demand for graduates with such expertise.[2] Girls and women, as well as people of colour, may be avid tech users, but they are not a significant part of the imagining and creation of code in Western nations. And it's not 'just' code. Code is used to process, describe, and shape knowledge, patterns, and systems – it is fundamental to many underlying rules that govern our world – so the diversity of its authorship has never been more important.

The problem lies in the history of the tech industry and the development of AI itself, which emerged almost exclusively in a white male context. Alan Turing's landmark 1950 paper 'Computing Machinery and Intelligence' opens with the compelling question, 'Can machines think?' – a question that has driven countless researchers and remains a cultural obsession.[3] By 1956, when the small group of all-white male mathematicians gathered at Dartmouth College to debate future directions in the relatively new field of computing, they christened this new idea 'artificial intelligence', and brought the idea of automating intelligence out of the realm of storytelling, into the forefront of science and culture. Few diverse voices have steered this conversation since.

Today, AI and ML are woven into our cultural and intellectual narratives and, even more concretely, into our daily lives, as devices such as Alexa and Roomba make their way into our homes. Indeed, from search engines to smartphones, technical systems shape our everyday experiences. The unseen challenge with AI technologies is the incredible potential for biased systems that exacerbate and extend social and structural inequities. No matter the field – be it economic, political, or artistic – AI and its intersection with the world is overwhelmingly male-dominated. Advances from Silicon Valley carry their own sets of biases that pervade technological development, such as ignoring the environmental destruction caused by the extraction of natural materials; enabling corporate structures and gig economies that hurt workers, not corporations; enabling racial profiling; enabling hate speech and algorithmically amplifying hate groups in social media; enabling the creation and dissemination of targeted fake news; and the manipulation of personal data for corporate interests. The list goes on.

Women and people of colour are left out of the AI conversation – in the making, structuring, and production of the code, and of the creations that AI can engender. Deep investigation is needed into the ways in which bias enters into AI and ML systems, especially as computing models change significantly over time. AI and ML systems are created by people, and therefore exacerbate and exaggerate human flaws; in particular, blind spots and biases as they relate to an intersection of identities: race, class, and gender. Thus, technologies such as AI reflect the cultures and values of their creators.

Today's ML, used to analyse large quantities of data, and the use of multi-layer frameworks of neural networks to perform deep learning, are vastly different computing models than the 'agent'-based model developed in classic AI projects. While critiques are emerging, these new computing models have profoundly shifted conceptions and abilities of AI. Further, the new models have promoted the fictional idea that through massive scale, decentralised processing, and iterative learning, computers are becoming 'more objective', and technology more trustworthy. Thus computers are learning, and even exaggerating, human biases that seem like objective truths.

The underrepresentation of women in tech has critical implications for the design and development of AI, because it is being designed and envisioned by a small elite with a particular perspective on the world. Research has long established that technologies are not, indeed, neutral,[4] and that computational systems can hold bias within. Bruno Latour famously noted that 'technology is society made durable'.[5] Technologies are in fact inherently political, as Langdon Winner suggests,[6] therefore *who* creates such technologies plays a key role in not only power but in our fundamental notions of intelligence, knowledge, and even consciousness. Philosopher Helen Nissenbaum has argued that new technology practices also bring with them social practices – norms are embedded into systems, encoding social expectations about how one should behave.[7] This can be seen in the erosion of the everyday right of privacy with emerging technologies.

There are few studies of contemporary AI systems that reveal their inherent biases. Safiya Umoja Noble's *Algorithms of Oppression* (2018) is an essential part of this new conversation, exposing how search engines reinforce racism. She examines the simple yet profound differences in search terms to show how, for example, searching 'white girls' returns vastly different results than searching for 'black girls', which produces results filled with pornographic sites and negative depictions of black women. Umoja Noble argues that a combination of corporate interests and monopoly-style control have created the condition wherein search engine algorithms privilege whiteness and discriminate against people of colour; in particular, women of colour.[8]

Apart from Umoja Noble, there has been limited feminist critique of the inner workings and knowledge models of AI since 1996, when science, technology, and cultural historian Alison Adam published *Artificial Knowing*,[9] setting the stage for thinking through the highly masculine epistemological model that AI was designed within. Both Adam and Suchman[10] showed that gender bias was encoded due to the era's programmed models for AIs, and argued that feminist thinkers should help shape AI futures. Since that time, the systems used for AI have grown exponentially more complex, with the use of neural networks and other novel knowledge models. Some neural networks are so complex that computer scientists do not yet understand how they are able to extrapolate outside their training sets. (As just one example, astrophysicist Shirley

Ho has found inexplicable situations where the training for an AI to model the universe is so intricate that computer scientists do not yet understand how these systems are functioning.) It does feel like science fiction to wonder how AIs are increasingly able to operate and formulate answers well beyond their 'input' training sets.

Gender bias enters AI in myriad ways. Training sets for AI and ML often contain skewed data, with training corpora considered 'neutral' often being anything but. Dominant search engines such as Google continue to return stereotyped and biased images (even after it was revealed in 2015 that an image-tagging app labelled a photo of African Americans as 'gorilla'). Such cases are even more important now that phenomena like racial and gender bias can not only be perpetuated but *amplified* by ML processes, as in the case of computer scientists who found that searching for women in a given data set produced an increasing number of women associated in searches for kitchens,[11] and that these biases were *exponentially compounded* as the algorithm continued to process. Thus, seemingly 'neutral' sources of data contain viewpoints, values, and epistemologies: something long recognised in domains such as the feminist critique of science[12] or psychology.[13] Indeed, the World Economic Forum has denounced the current situation with gender and AI as 'failing the next generation of women'.[14] Biases emerging or being exacerbated with AI and ML include selection bias, as the data is selected to train AI; the illusion of a machine's objectivity; lack of transparency within data sets; and the partial representation of reality, or perspective, in the data collections we use.

AI and ML urgently need to be examined from an intersectional feminist perspective.[15] Creating a standpoint with key intersections around gender is a crucial way forward to identify AI's possibilities and limitations. Indeed, if only 12 per cent of ML researchers are women, even fewer scholars studying the field will have the technological vocabulary and understanding of such systems.[16] Injecting a feminist perspective into a hotly contested field that has grown up largely in the hands of male scientists has the potential to galvanise important discussions and research into how AI might not only make room for women and underrepresented groups, but enlarge, improve, and render its structure and practice more ethical as it continues to pervade our minds, work, and world.

As society begins to attend to the pressing questions of automation, ML, and the boundaries of the human, we must examine the systems we use to represent this information.[17] We need to test out the key assumptions that are 'taken for granted' and the prevalent myths in computational intelligence, for example, that AI makes better decisions than people, that AI cannot be regulated, that mechanic processes are more equitable than human ones, and that AI's lack of empathy is its strength.

Steps are being taken, such as the Turing Institute's focus on redressing gender inequality in AI and data science fields at the policy level. At the conceptual level, the tech community needs to understand that biases are pervasive and must be designed to actively counter such bias.

Rather than accept AI systems as is, or as 'truths', we need to understand, through a feminist lens, what it might mean to rework these norms, and prioritise new models in which women and people of colour also steer the conversation. Instead of 'adapting' to technological change in a passive way, we must actively shape how it works, what logics are prioritised, and what values to insist upon, so that the collective good is reflected in code.

8

Colouring Outside the Lines: Constructing Racial Identity in Intelligent Machines

Anita Chandran

Recently, when I open a dating app (a rare and painful thing), I find myself idly scrolling through my own profile, trying to imagine what others see. As a British Indian, I wonder if I am viewed as a whole person. I am sure that almost everyone who uses dating apps has similar thoughts. For a person of colour, it is different.

Are people making assumptions about me based on my ethnicity? I ask myself. *Perhaps they are choosing to challenge their own prejudices? And is this better?* I wait to discern how others construct my racial identity, how to adapt my behaviour to the audience. *How much can I be my real self? Do I feel compelled to conceal aspects of my upbringing or culture around certain people?* The very sight of my skin colour essentialises me; it holds me in like a seat belt. I wonder how to get out of the car.

Intelligent machines are subject to racial profiling. How they are profiled will depend on their form (embodied or disembodied) and who is interacting with them – the dating-app scroller – the person who constructs their racial identity. Mostly, this role belongs to humans: the developers and designers of intelligent machines, as well as the people who interact with them on a personal or professional basis. It is the consequences that result from these human–artificial intelligence (AI) interactions that concern us here: the overwhelming 'whiteness' of AI, how racial prejudice and fetishes are projected on to intelligent machines, and the negative impacts on people of colour.

Just as with online dating, when we interact with AIs, we project our biases and preferences on to them. We make assumptions based on the way they look or act and these assumptions are influenced by the form they take – whether they are embodied (with a physical form such as a robot) or disembodied (such as a voice assistant or chatbot). We assign racial characteristics to them instinctively, just as we racially profile one another.

Humanoid robots have 'skin', facial features, and hair. Some take human form – they are bipedal, with increasingly humanoid faces. Disembodied AIs such as Amazon's Alexa are given names that act as cultural signifiers, and voices with 'friendly' dialects to make them seem approachable. This verisimilitude extends to their names, voices, accents, gestures, and manners. Think of the dulcet tones of Amazon's Alexa, or Scarlett Johansson's disembodied voice in Spike Jonze's *Her*.[1] The design features that enable AIs to be 'more human' also assign racial identity. Indeed, even when given the option to abstain, people tend to ascribe a race to robots based on the colour of their skin,[2] which suggests that robots are exposed to the same structures of racial discrimination as humans.

AIs are generally developed as white, particularly in physical appearance.[3] This is in part due to the 'monopoly that whiteness has over the norm', both in researchers working in AI, and in portrayals of intelligent machines in fiction and film.[4] Facial features are predominantly modelled around traditional, Western (read: white) beauty standards.[5] Just like dating in the Western marketplace, value is bestowed upon whiteness and anything different must justify the space it occupies.

The stereotype of 'Man' as white, able-bodied, cis-gender, male, and middle class[6] is mirrored in research and development teams in AI. For example, a tiny 2.6 per cent of Google's full-time employees are Black, and 3.6 per cent Latine.[7] Indeed, many technologists would prefer to describe themselves as 'colour-blind', believing that they are adopting a stance of racial neutrality in development.[8] But society is not racially neutral, and racial literacy – an understanding of racial power dynamics – is vital if AI systems are to be used by people of varying ethnicities across the globe.

The creation of by-default 'white' intelligent machines excludes non-white users. In some existing technologies, for example, non-white English dialects and accents (in particular, certain African American dialects) suffer in the programming of speech-recognition software. This is exacerbated by racial bias in marketing,[9] with certain ethnic groups being seen as 'outside the target market for AI', but also by the lack of diversity in technology companies. The choice to use non-white characteristics (voices, names, dialects) is often considered controversial, and even non-white developers may choose not to rock the boat.

But the diversification of intelligent machines also requires careful handling. Since AI is predominantly designed and developed around the preferences of white people, diversity may simply be tacked on to an already white structure, like draping a scarf on a mannequin. Superficial adjustments such as changing skin tone and accent do not address the issue of how AI meets the needs of non-white people. This can result in quick fixes, such as 'ethnic' traits being added in the form of lazy stereotypes. A software-developer colleague once said to me, 'racial identity often feels like it's being added in post[-production].' This does not make AI more accessible. Instead, it alienates non-white people, meaning we must adapt our behaviour to use these technologies, shedding the markers of our identity to be allowed into the future.

Once race becomes an identifying feature of intelligent machines, certain harms may accrue. We already see that humans treat AIs better when they perceive them to share the same racial identity as they do.[10] It is quite likely that robots with non-white cultural signifiers will be subjected to the same dehumanisation that non-white people face in society today.[11]

The potential for racial abuse is already reflected in the treatment of AI assistants such as Apple's Siri or Amazon's Alexa. Although these assistants are ostensibly 'genderless', they are usually given a feminine voice, prompting users to gender them as women. Research confirms that AI assistants are subjected to large volumes of sexual harassment and verbal abuse.[12] More harrowing still is that they are programmed to be deferential, humble, or even flirtatious when rebuffing advances from users.[13] People interacting with AI assistants are often rude and commanding, refuting 'please' and 'thank you', and demanding a high degree of subservience – an expectation that bleeds through into society.[14] In fact, research from Harvard University suggests that the treatment of AI assistants will detrimentally affect the way women are spoken to, conflating womanhood with 'servitude'.[15] This is an alarming outcome that could easily be mirrored in AIs with different ethnic markers.

One particularly insidious instance where racial prejudice may creep into AIs is the advent of the sexbot, robots designed for use in sexual activity whose users are predominantly men. Many are physically embodied, with some being equipped to take a more active role (e.g. Synthea Amatus's Samantha, can 'refuse' sexual intercourse[16] – which raises a fascinating question on a robot's ability to give enthusiastic, authentic consent – a subject for a separate essay).[17]

For many people, ethnicity is a sexual fetish, rooted in deeply disturbing assumptions: for example, the stereotype of Asian women as 'demure and submissive', or Black women as 'domineering and hypersexual'.[18] Such fetishisation is prominent in pornography and, to some extent, in racist dating app algorithms.[19] People may argue that their preferences for certain ethnicities are legitimate, but they exist alongside society's prejudice and racial stereotypes.[20]

No one is colour blind. Racial fetishes such as 'I just don't find X people attractive' or 'I only date Y people' are steeped in a subtle conditioning that begins in childhood and continues throughout life.[21] They stem from the same bias that keeps Black people out of higher-level jobs, while promoting Asian people in technical fields. This bias influences the way our preferences form, making it impossible to unpick one from the other.

There is a significant commercial market for sexbots that caters to and preserves racial fetishes by aggregating characteristics that fill stereotypical assumptions about minority ethnic groups. Their existence creates a world where it is possible to have a sexually intimate connection with a pale (and offensive) imitation of a non-white person. This reinforces dangerous tropes relating to the sexual behaviour of people of different ethnicities, and also stops users from having to interact with real people of colour and interrogate their problematic assumptions. As with the sexual harassment of AI assistants and chatbots, sexbots can also be abused by their owners promoting harmful behaviours and enforcing an insidious imbalance of power during sex.

Sex is a humanising act combining physical touch and emotional intimacy. Sexbots will occupy significant roles in the lives of some of their owners, potentially replacing the need for relationships with non-white people. Having a sexbot that fulfils your racial fetishes can feel much less complicated than actually being with a non-white person, and/or having to explain an interracial relationship to your peers. It also allows for interracial relationships to be something that can be hidden away as shameful secret.

This also leaves space for the existence of racialised machines in spheres where it is considered acceptable to have active racial prejudices. These are situations where a person can expend their frustrations on a non-white AI. Intent matters – if you racially abuse your AI, what happens when you switch off and enter the real world? You continue to be racist, and you become empowered to be so. And others see these interactions, in the same way that children hear adults berating Siri or Alexa, and learn that they are acceptable. This bleeds through into the treatment of people of colour every day.

Navigating life as a non-white person in a predominantly white society is already complex. Once, on a date, the person I was with blurted out 'your English is amazing'. I have had to listen to descriptions of my skin ('chocolate', 'caramel', 'coffee', 'toast'). I've been offered judgements about my career ('oh, I just assumed your PhD was in medicine, don't know why!') and heard frankly unimaginative adjectives to describe my style ('spicy'). The complications of social stigma, racism, and fetishisation weigh heavily on people of colour.

AI has the potential to permeate every aspect of our lives, and the power to influence our behaviour and culture in a lasting way. It is vital that its development abandons the norm of whiteness and commits to understanding and prioritising diversity in design, and racial literacy in its developers. If it doesn't, we run the risk of alienating non-white people, and reinforcing the prejudices we already face on a daily basis. The seat belt pulls ever tighter.

9

Never Love a Robot: Romantic Companions and the Principle of Transparency

Joanna J. Bryson

Transparency is considered a desirable quality in most relationships, but does this hold true for love? While we all agree that intimate relationships must be based on firm foundations, some people consider too much transparency or openness a threat to that firmness, given human fallibility. Evolution has shown that we must mythologise our core relationships. Our hormones alone seem to overwhelm pragmatic assessment of our partner's true quality.[1] Good reasons have been offered: partner quality is difficult to assess, and changing partners is very costly, so we should be slow to commit *and* also slow to violate a commitment, because care before a commitment means we are unlikely to do better if we try to change.[2] The costs and risks of some sufficiently bad relationships may be so overwhelming that their termination is worth the costs and hazards of moving on. But it may also be that a bad relationship can be improved through increased transparency on both sides, and an adjustment of our expectations and responsibilities.

Excessive transparency is also viewed as a hazard in two other domains: digital technology and governance. I would argue that the resulting opacity has also resulted in abuses. First, in the case of digital technology and particularly artificial intelligence (AI) and robotics. Here the design goal has always been to make a system 'transparent', in the sense of *invisible*. We developers are told we should empower our users, rather than make them aware of how the system actually works. Second, and similarly, in governance, no one wants to know how the sausage gets made. Governments and consultancies alike are often advised or persuaded to make their citizens or clients look good – to give away credit and then stay out of the way, while providing the infrastructure for a vital society. Consultancies, however, must demonstrate their value regularly, when they present their invoices and attempt to get paid. Perhaps the coercive power of the state, in contrast, now ironically reduces itself. Too many voters are too easily persuaded that the present order in liberal democracies is giving them literally nothing in exchange for their taxes.

The problem with transparency in this invisible sense is that it becomes difficult to assess costs and benefits. Turning again to technology, we find it almost impossible to know how many of our devices have gathered how much data about our predilections, problems, debts, savings, joys, fears, or desires, nor what can be done with that information. What advantages are we receiving, at what cost? How can we know, when both are hidden? Likewise, few appreciate or understand how much they are enabled – empowered – by the rule of law: instead they chafe against regulation as a restriction to their liberty, and see no reason to pay the taxes that sustain their way of life.

Transparency in AI and Other Digital Artefacts

The requirement for transparency in AI has now been incorporated into the policy of the Organisation for Economic Co-Operation and Development (OECD), including their five core 'Principles of AI', also now ratified by the G20. Transparency forms a triumvirate with accountability and responsibility to maintain justice and social coherence in the digital age. Humans must be held responsible for the technology they develop, own, and operate. They are only responsible to the extent they can be held accountable. They can only be held accountable if the systems in which AI is implemented and through which it is operated are transparent. It follows that we must, at least in theory, have access to information about what a system 'needs' or 'wants' for its successful operation, why it behaves as it does, and so forth.

The framers of the original five-principle system for AI ethics (and the first national-level soft law in this area) went beyond conventional legal concerns into the area of human well-being.[3] Deception was a primary concern of the authors of the UK's 'Principles of Robotics': that users should not be deceived into over-investment of time, love, or any other resource into AI by a falsely anthropomorphic or zoomorphic appearance or behaviour.[4]

Such considerations are, of course, antithetical to the marketing of an AI romantic companion. If we assume consumers want to go beyond mere physical stimulation, that they want romance from an artefact implies a desire for reciprocated animal attraction. Nothing mechanical will experience the viscera of biological sensations or drives, at least not to a degree of similarity that other mammals that we do not find remotely romantic. But does love require reciprocation?

Love and Bonding as Biological Processes

In all species of mammal, most birds, and even some fish, parents are deeply affiliated with their children, emotionally bound to throw themselves in harm's way to protect them, and apparently often revelling in their presence. Evolution is quite likely to select for individuals who take pleasure in the well-being of their offspring. Pregnancy, childbirth, and even copulation can be extremely hazardous, and ultimately result in producing competitors for the same resources, in the same locations, as the parents. Given all that, the processes leading to parenting must be deeply appealing, or somehow otherwise inevitable, in any species. Perpetuation is the core of biology.

Pair bonding between those who produce children – and even those that do not – is rarer than bonding with offspring (i.e. not ubiquitous), but still hardly unique to humans. In species such as birds, and even some insects, where two animals are required to adequately provide for their young, two individuals tend to do this together, and often (though, of course, not always) those two tend also to be the two parents of all or at least most of the offspring. Sometimes older children or unrelated 'helpers' join in the childrearing. In the latter case, these unrelated helpers often have a chance – if they survive long enough – to become one of the collective's lucky parents. Examples of this sort of family system are the mongoose and the clownfish. Clownfish have a particularly interesting experience, since whichever of the group is the largest gets to express as being female, which often involves a single individual playing more than one, not just gender, but sex, over time.

Mammals like us don't usually fall into the category of pair-bonders, but that doesn't mean we don't have a lot of sex.

In many species, both homosexual and heterosexual sex is practised that is unlikely or impossible to lead to children. Sex still provides something, though – evidence of commitment to a relationship, evidence of trust, or at least a willingness to enter into a vulnerable position with a partner, and information concerning current physical status. Such information can be important not only for choosing the parents of your offspring, but also for choosing coalition or hunting partners.[5]

Humans are one of the very few species of large mammals, and indeed of primates, that do pair-bond. Other pair-bonding primates include gibbons and the birdlike callitrichids. We can tell when in our evolutionary past we (at least females[6]) became more monogamous, because testicles then didn't need to be as large. Males of other types of chimpanzees than recent hominids practise a lot more sperm competition to try to ensure their own paternity, and this turns up in fossils. Personally, I'd like to believe that as we moved out from the jungles to the savannah, we became more romantic and devoted as couples in response to an environment that required co-parenting, but then of course I'm female. Even in chimpanzees, though, we do sometimes see long-term and apparently devoted relationships between two individuals that extend for years, particularly among young adults.

So, does it matter that, for many of us, the relationship we place at the core of our existence, that we organise our lives around, is essentially a mythologised hormonal trip designed to perpetuate our species and ourselves through a stable social group? In my opinion it *should* not matter, because this is what humans are. What we are is a *fact*, and these other facts about us can help us know how we can find comfort and individually flourish. Whether it *does* matter or not will vary by the individual. It will also vary with the resiliency of our culture and the various arts that give us identity as we come to better understand ourselves.

Never Love a Robot

So what does all this say about loving robots? It says that humans are pre-adapted to bond, and so we will easily bond with objects we are – by some combination of biological and cultural predisposition – predisposed to admire in that way. But, just as with humans, it is clear that not every relationship with a robot is worth getting into. We must remember that robots are commercial products. Robots, whether humanoid or not, will have cameras and microphones, and possibly other sensors, perhaps concealed. Their use case – to serve as a surrogate lover – requires storing information about you, very intimate and sensitive information, such as your health or emotional state. If a robot with such data ever, ever, ever connects to any network, that information becomes available to its makers, its owners/operators (if that is someone other than you), and even to arbitrary hackers and blackmailers. Without adequate transparency and regulatory oversight, we will be unlikely to know whether a robot ever does connect to a network, or even, in the worst case, to the open Internet.

So, what about robots – will they love us? With AI, we play the role of evolution. We select and design devices not to perpetuate, but to serve our needs or desires. We can therefore choose to make robots as monogamous, devoted, obsessive, caring – even violent – as we wish. As with every other phenomenology, such bonds are not perceived by any machine with anything like the viscera a human would experience. Machines have entirely different sensors and representations than we do. No machine will ever have as much chance of direct empathy for us as we have for insects, and nowhere near as much as we have for reptiles. Of course, machines can be and already are being used to fake human empathy by simply looking at data about humans and anticipating appropriate next actions.

They say you should never say 'love' to a tennis player, because to tennis players 'love' means nothing. When it comes to AI and robots, love means whatever someone or some corporation has programmed it to mean. Even if a robot or other robot system is programmed socially, for example through learning from text or spoken language, and winds up with the average meaning of most language producers in the language concerned, as a digital artefact that meaning can be easily overwritten or overridden. Loving a robot is far more dangerous than loving a tennis player, because there will be no reciprocity, and the nature of the relationship will always be subject to immediate change whenever the robot has contact with a network, or even a hacker in your own home. To a robot, 'love' can mean anything.

10

Can Robots Be Moral Agents?

Amanda Sharkey

What would it mean for a robot to be a moral agent? A moral, or ethical, agent is one capable of acting rightly or wrongly that can be held responsible for their actions. A full ethical agent can make ethical judgements and justify them.[1] A human with consciousness and intentionality is a full ethical agent. But could a robot be one?

This question matters because of the likelihood of placing robots in situations in which their actions could have a significant impact on human beings. For instance, a social robot deployed in classrooms might have to distinguish between children's good and bad behaviour.[2] A care robot installed in the home of someone with dementia might 'decide' not to let them leave the house, or to report concerns about their health or behaviour.[3] A robot used for crowd control by the police might be required to spot delinquent behaviour. In warfare, autonomous weapons are being developed that could be used to make decisions about who to kill.[4] Do we want to place robots in positions of authority if they do not have a sense of what is right or wrong?

It is sometimes suggested that robots could make better decisions than humans because they will not be biased, nor influenced by prejudice, racism, or anger. For instance, Ron Arkin, a professor at Georgia Institute of Technology, has claimed that a robot soldier could be more moral than a human soldier because it would not become angry or seek revenge.[5] When humans make decisions, they can be influenced by their own emotions and may fail to choose the outcome that leads to the maximum benefit for the greatest number of people. A programmed robot on the other hand, or a programmed computer algorithm, will not be distracted by its own preferences, or its own opinions, or so the argument goes. A robot in a classroom will not have favourites or be racially biased against some of the pupils. A care robot in the house of a person with dementia will never get impatient about constantly being asked the same questions.

However, humans tend to overly trust the outputs and actions of robots and computers. Perhaps some of this stems from our early reliance on calculators. If we add up a set of numbers in our head and then check the result using a calculator, we do not think that the calculator could be wrong if the answers differ. But as we move away from such straightforward examples, there are fewer grounds to believe that a computer, or robot, is always going to be right.

An example of inappropriate belief in the output of a computer program, despite considerable evidence to the contrary, can be found in the UK's Horizon post office scandal.[6] The Horizon software installed in post offices made errors that led to shortfalls of thousands of pounds. Sub-postmasters were held responsible, and many were convicted of fraud or theft, became bankrupt, imprisoned, or lost their jobs. It was assumed that the Horizon system could not be wrong, but in 2019 a High Court judge ruled that the information technology (IT) system was not 'remotely robust'. The case, still being settled, illustrates some of the problems and risks that can result from an over-reliance on a computer system.

There is a growing awareness of the extent to which computer algorithms can be biased. Even so, the problem of inappropriate trust can be amplified when a computer program takes on the embodied form of a robot. Robots can be created with a variety of appearances. A robot intended as a companion for an older person could be designed to look like a furry pet. Or it might be a humanoid with a friendly face and voice. When a robot is built to resemble a human, with an emotionally expressive face, people are likely to expect it to be able to do and understand more than it can. This expectation is strengthened when the robot is equipped with functions such as face recognition, emotion recognition, speech recognition, and speech production. All of these can create the impression that the robot has emotions or even that it is sentient. Added to this is the human tendency to be anthropomorphic, and to imagine that even inanimate objects such as cars have personalities and emotions.

But, whatever the appearance of a robot, however friendly, human, or animal-like its demeanour, it is still a programmed artefact. As such, it can neither be held responsible for any decisions it makes, nor does it need to be protected from suffering. A robot cannot be held responsible for its actions or outputs (in other words, it is not a moral agent), and it does not need to be given rights or protected from harm (it is not a moral patient). People may amuse themselves by treating it anthropomorphically as if it was their friend or a living entity, or even believe it to be so, but in reality, it is neither. There is no need to be concerned about the welfare of a robot pet or companion, because it has no feelings. If a robot was to act with cruelty towards a human, it cannot be held responsible.

Let us now consider the reasons for claiming that robots are not moral agents or moral patients, that they should not be trusted to make decisions that have a significant effect on humans, and that they are not able to care about humans or understand them. Essentially, they boil down to the observation that they are not made of the right kind of stuff. Robots are machines that have been assembled and programmed by humans. As Ada Lovelace once said, a program (or robot) can only do 'what we know how to order it to perform'. Programmed machines are entirely different to living ones.

Living machines, from bacteria to humans, have evolved to be what they are as a result of the interaction between their physical form and the environment. This evolution has not required the intervention of a human programmer. The body of a living creature is an integrated whole, in contrast to the body of a robot that is composed of parts, any one of which could be removed without pain or suffering.[7] There are debates about what the minimal requirement is for conscious experience and sentience, and whether, for instance, fish can feel pain (there is some evidence that they can). Humans are both conscious and sentient, and even though their cognition differs from ours, it is generally accepted that other mammals are too. Humans and other mammals feel pain, and also experience pain and suffering when their loved ones suffer.

Patricia Churchland has convincingly argued that morality is grounded in biology.[8] Humans and other mammals extend their self-maintenance and avoidance of pain to concern about their immediate kin. They form attachments to them, and these emotions form the basis for more complex social relationships as they seek and are rewarded by approval and belonging. This leads to the internalisation of social standards, and the emergence of morality: rudimentary in animals such as dogs, more sophisticated in humans. Humans have the further ability to reflect on their ethical decisions, to make moral judgements about others, and to develop their own moral sense about how they, and others, should behave.

A robot, on the other hand, is dependent on humans for its development, and its programming or training. Its actions can have an effect on other robots, or on humans, but that does not mean that it is able to care about any consequences. It might be programmed to look after humans, but it could equally well be programmed to harm them. A robot does not even care about itself, let alone extend that concern to other people or try to please them and fit in with social norms. Although a robot is a machine, it is not a *living* machine, therefore it cannot suffer. It would make no sense to offer pain relief to a robot, in contrast to the beneficial effects of providing analgesics to suffering living creatures.

Some counter arguments to this position have been made by philosophers. It has been argued that the moral status of robots should not be based on what the robot is made of or capable of. David Gunkel suggests that robots may deserve rights.[9] His claim is that if we take what he terms a 'relational turn', what matters is how we relate to them. If we behave towards a robot as if it were an entity with which we can have a meaningful relationship, then it deserves rights. John Danaher[10] makes a related argument, that if a robot behaves in a way that closely resembles the behaviour of something that we recognise as having moral agency, then we should take a behaviourist stance and assume that the robot has the same agency. The problem with these abstract, theoretical, arguments about futuristic robots, is that they distract us from the real risk that the deceptive appearance of current robots will lead us to over-estimate what they can do, and to place them in roles that are inappropriate.

There are those who argue that if robots are to be used in sensitive tasks, we should try to make sure that they behave ethically. There are two main approaches to doing so: (1) providing them with a set of moral and ethical rules to apply, and (2) training them with examples of moral behaviour so that they can infer a set of moral and ethical rules.[11] Both approaches have so far met with limited success.

The problem with the provision of a set of moral rules is that, unless they apply to a very limited and constrained domain, they require interpretation. Any such interpretation requires an understanding of the human social world that a robot is not going to have. The science fiction writer, Isaac Asimov, proposed the three laws of robotics:[12]

1. A robot may not injure a human being or, through inaction, allow a human being to come to harm.
2. A robot must obey the orders given it by human beings except where such orders would conflict with the first law.
3. A robot must protect its own existence as long as such pro-tection does not conflict with the first or second laws.

Although these rules are fictional, a consideration of how they might be implemented in code reveals some of the problems of programming a set of moral rules. There are many ways in which humans could be harmed by a robot – from physical damage to psychological trauma and emotional upset. As well as the problems of programming a robot to understand a human's commands, there is no easy way of programming it to ensure that its actions do/will not lead to harm. How could the programmer anticipate all the possible effects of a robot's behaviour?

Humans develop their morality on the basis of their experiences and interactions in the world, but that is because they are part of that world, and because they care about other people and their opinions. It is possible to use machine learning and many training examples to develop a computational arte-fact that can, for example, play the game Go, and even beat the world's greatest player at the game.[13] This is because the domain of Go is a specific and constrained one. The open-ended domain of all moral decisions is quite different. It is diffi-cult to demonstrate in principle that carefully selected training examples could not lead to the inference of a set of moral rules. But so far, there is little evidence that machine learning can be used to create a robot that can make moral decisions in unanticipated situations.[14]

In any case, attempts to create artificial moral agents depend on human intervention, since humans choose and program the set of moral rules and select the set of training examples. Since robots rely on humans for their development, it should be clear that humans remain responsible for their actions, even if the robot behaves in an unexpected and unpre-dictable manner.

We cannot abdicate responsibility for robot actions to the robots themselves because they are not moral agents and cannot be held accountable. Therefore, they should not be used, without meaningful human control, in domains where their actions could have significant impacts on humans. Robots cannot care about humans and the effects their behaviour might have. Nor can they understand, or be programmed or trained to understand, human behaviour or communication. Robots also are not moral patients: they cannot suffer. Any imagined reciprocal relationship with a robot is just that – imaginary. In conclusion, I propose that instead of worrying about robot rights, we should be concerned instead about the rights of animals and creatures that can suffer, and about the rights of humans to be protected from arbitrary statistically based decisions made by robots and computational algorithms.

11

Words to Life: Creativity in Action

Peter R. N. Childs

I woke up and there they were. I couldn't believe it for a moment, but I was so pleased. The rhomboid massive earrings I had just dreamt about and seen myself wearing. The sensors in the pillow had done their thing, streaming data to the bedside transponder, thence to be analysed, connections and interpretations made, and an instruction sent to the household three-dimensional printer. Resplendent aubergine and orange colours in lacquer on a slender silver structure. I know the base material is some kind of sintered silver material, not forged or cast. However, it's not the tacky trashy plastics of the early object printers. My domestic robot will have picked up the earrings from the printer hopper and arranged them with my clothes, selected to match of course, for the morning. This is just part of the life management algorithm at work for me. Settings selected so I have some control – I do have my limits but it's just so exciting sometimes for your dreams to be realised so quickly. Oh, they do look good – the earrings, that is.

We dream and think; we speak and write. Every word can be transcribed. Every word analysed and connections made, whether intentional or otherwise, whether under our control or just because someone or some system somewhere is running an algorithm that is exploring a database. But to what end? This scenario is an example of subconscious and consciousness fusion at work, all enabled by the algorithm I have subscribed to. It uses all sorts of artificial intelligence (AI) such as machine learning, reinforcement learning, and neurolinguistic understanding, coupled with sensors and physical interfaces such as machine vision, three-dimensional printers, and robots in order to do its thing – giving me what I want sometimes even before I quite realise it.

Creativity and innovation are often associated. Some practitioners and commentators choose to regard creativity as imagination with responsibility. From a societal perspective, we tend to want things that are useful or serve a purpose and have spent much effort on such evaluations, exploring values associated with pleasure to technical function. AI has matured to the extent where we can readily apply its algorithms and processes to generative activities. Indeed, AI and creativity are now interlinked, with many examples of generative algorithms producing examples of art, dialogue, and music, where it can be difficult to discern its origin. This chapter explores the blurring of that which originates from a human and that from other mediums in a world where co-creation is becoming the norm. From knowledge to new knowledge and new creations enabled by a fusion of human thought, memory, and inventiveness intertwined – and now let loose in the form of learning algorithms that the originators barely recognise or comprehend – our world is indeed experiencing change at speed.

The example at the start of this chapter could be considered as facile or self-indulgent. Such a view may depend on whether you like and value earrings and what your subconscious gets up to. How about if the same suite of technologies is applied to another type of application? We have a long history of making enzymes, for example, by stringing together hundreds of amino acids in a specific manner. We also have advanced capabilities in nucleic acids, proteins, carbohydrates, and lipids. As a result, we now have the capability to produce artificial cells, sometimes regarded as a building block for living entities. While there is no universally accepted definition for life, many observers will include notions of develops, defends, divides. For the more advanced forms we can add devises and designs – and, of course, death. So, if we can produce enzymes and proteins and other building blocks for cells, some of which can already be judged to have passed the classic Turing test, is it beyond the realm of possibility that we might be able to produce life forms and even advanced life forms?

I have always enjoyed talking to my dad. The twists and turns of a loving relationship and interesting outlook have been a great blessing. I have the privilege of working in a university where the students and staff are constantly challenging the notions and limits of creativity and innovation. This has resulted in all sorts of inventions, gadgets, and gizmos. I often talk to Dad about these with great enthusiasm and have learned to pick up on his wry smile. So I ask, 'OK, I know you like the idea, but what are you actually thinking?' And then I wait for the kindly response that I know will come. One of the courses I run is called 'Billion-Dollar Question'. The notion is that if you want to be a billionaire you need a big question, and if you want to be a social entrepreneur you may need an even bigger fund in order to pay for the social interventions. The student outcomes have been incredibly diverse, from the questionable animal welfare last-mile delivery system using dogs and technology, to charging for fresh air in your neighbourhood enabled by major infrastructure and plant life management, to mention but a couple of ideas. One team responded to a lecture I gave on what it would take to produce intelligent life, by proposing a new form of pet. I took great pleasure in recounting the idea and the suggested pet form to Dad. But there was that wry smile and out it came, that decisive verdict, 'It's going to take a lot to improve on a cat.'

Evidently, we have a long way to go before we can claim with confidence the capacity to produce intelligent life forms. Our integrated software and hardware systems, however, have elements that could be construed to offer aspects of the capability to develop, defend, divide, devise, design, and perhaps even demise. We all have experiences, including many frustrating ones, of interactions with chatbots. These systems provide us with a foretaste of the emerging capability of augmented systems, with humans often providing a fill-in function, such as packing or delivery, until this too can be designed out. However, it may be that a human in the loop route provides added value and that, in the end, this is the way things turn out.

Electroencephalogram (EEG) and functional magnetic resonance imaging (FMRI) are technologies commonly used in medicine that can also be applied to imaging what a person is thinking. Going back to our what-if scenario at the start of this chapter, use of EEG has already been demonstrated to enable what we are thinking about to be visualised. This work remains at an early stage and with limited application so far but we have been able to use a bonnet with 128 EEG sensors and with a few seconds of data, following initial training of the data base and system, to enable 64 x 64 pixel images of what we are thinking about to be displayed.[1] The software in this case uses a machine-learning framework called a generative adversarial network in order to discover more optimum ways of analysing data. The results have been impressive with good discrimination between images and recognisable forms being produced. Already the work is being explored as a means of enabling a person to communicate the form of something they are thinking about, such as a gift they might want to give to a relation or friend. The approach could also be used for human-in-the-loop design, where potential users and stakeholders just need to imagine a shape or form for a product or design. This image can then be used to inform the more detailed work of a design or business development team or algorithm to produce the design – or even to allow existing products to be searched to see if there are purchase match options that can be provided.

Creativity is often regarded as the ability to imagine or invent something new that is of value. For a long time, creativity has been limited to a capacity associated with humans alone. It may be possible to challenge such a notion, with, for example animal behaviours such as crows and rooks finding, transporting, and then using diverse forms of stick in order to construct their nests, or predators devising new approaches to trap prey. There may also be some precedent of creativity described in chapter 1 of Genesis in the Bible; a chapter that could also help inform a definition for life. The explorations here, be it intelligent systems for producing products from our subconscious, or new life forms devised from our ever-advancing capabilities in the synthesis of complex molecules and structures, represent a new paradigm where we have become little gods – where what we think or say comes into existence. We have the potential for our thoughts to be formed, monitored, possibly even adapted and modified, interpreted, augmented, and executed through brain machine interfaces and associated systems. Whether humans remain as agents in control or subjects will be interesting to see unfold.

12

An Anthropology and Philosophy of the Future

Richard Watson

Dear reader,

You have just read a novel about our technological seduction and the future of human love,[1] and so I expect you may be wondering what to make of it all. How might you approach such silicon (and silicone) speculation, and the future more generally?

I've been using the F-word since 2004, but it wasn't until 2007, when I wrote a book called *Future Files: A Brief History of the Next 50 Years*, that people began using the F-word to describe me.[2]

One frustrated reader asked: 'What kind of futurist are you … reluctant?' I took this as a compliment. This reader had a singular technological future in mind, one that he believed was inevitable. I disagreed. I'm not saying he was wrong, and he didn't tie his vision to a particular date, but I collect historical books about the future and the more I look backwards, the more apparent it becomes that things rarely turn out as we expect. Perhaps that's why historians make such good futurists. They're not technological determinists and do not commonly make the mistake of confusing rapid movement with enduring progress. They've seen it all before, history laid out like an ancient land, where the fresh ink of new ideas stands out in stark relief against the enduring nature of basic human needs.

What I'm getting at is that the future is not immoveable. Rather, it is a distant and hazy land open to investigation, a terrain that is endlessly being remade by human action, inaction and, most of all perhaps, reaction. This often means a series of confident advances followed by a number of hurried retreats. I recall a conversation in the early 2000s with a number of fellow futurists, most of whom were convinced that e-books were the future – paper books would soon be extinct, and the numbers proved it. But using historical numbers to predict future direction can get you into a heap of trouble. Trends, in particular, are problematic over the longer term because the world isn't binary, it's systemic. Trends can bend, overlap, and sometimes snap.

Fashion is full of examples, or one might mention 3D television, robotic pets, open borders, or net neutrality. Random events and second-order effects also interact in highly complex and unforeseeable ways. We should also be careful not to write off things simply because they are old. Many technologies have endured precisely because they are so good, and paper is a prime example. Over time, people have worked out that paper and pixels each have their own advantages and disadvantages, which suggests to me, at least, that the future is often *and*, not *either/or*.

Another pertinent point, and one that you may have experienced reading this book, is that the future has an asymmetric distribution. This is certainly true with sex robots, which

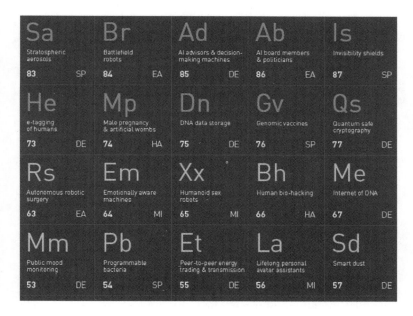

Figure 12.1 Detail from 'Table of Disruptive Technologies' (Watson and Cupani, Technology Foresight Practice, Imperial College London, 2018). The table ranks 100 emerging technologies against impact and time to ubiquity. Note 'sex robots', centre.

already inhabit the minds, homes, and workplaces of a handful of people, but not others. The future generally kicks off around the fringes, often with an idea that infects younger people or takes hold in certain locations, before it sometimes spreads out and becomes ubiquitous. You can therefore discover fragments of the future in research laboratories, start-up companies, or lurking beneath your feet in certain geographies. Japan, for example, has emerged a leader in robotics, especially robotic companions in care homes and robotic assistants in kindergartens, for decades. This is partly because of Japan's rapidly ageing population and low birth rate, and partly because Japan is arguably ahead of anywhere when it comes to societal acceptance of robots. In Europe and the United States, robots are generally seen as destroyers, while in Japan they are usually seen as saviours.

Similar cultural bifurcations exist in attitudes to privacy and surveillance, for example. If you're trying to be a better futurist, a consideration of cultural context and psychology is important – not everything works everywhere, or for everyone. Our world is a system and we cannot consider anything in blissful isolation. We must test any prediction in the context of broader factors such as economics, the environment, politics, and societal values. For instance, the future of robotics should be considered in the context of trust, regulation, pricing, security, the environment, and so on. Doing this won't guarantee we'll be right, but it should ensure we'll be less wrong.

Over the years, I have found that, overall, the best approach is to proceed with caution, to keep an open mind about what's possible, and to alter it when circumstances change. This is easier said than done, of course, because our worldviews, along with our cognitive biases, mean that what we think we see clearly is often distorted.

But why do we care about the future anyway? I think the answer is that humans have always been interested in what lies over the next hill, or around the next bend. This probably stems from a need to anticipate danger, or locate new opportunities, although a more modern explanation might involve sense-making. One of the things I've noticed when giving talks to audiences around the world is that while some people's eyes glaze over when I start talking about the future, others sit up and can't wait to ask a question.

Writing this piece in the middle of the COVID-19 pandemic, I imagine that once the virus has withered, or we've learnt to live with it, people may even pay a premium to interact physically with others, because I believe there's a deep and unchanging human need to connect physically and to share stories in person. Zoom, Skype, and other digital alternatives are fine, but they are somehow sterile and miss out on so many layers of human connection and communication.

As for why else there's so much curiosity about the future, and why institutions engage the services of futurists like me, I suspect it's partly because they think that glimpsing an outline of the future will enable them to ride waves of opportunity. They may also be attempting to minimise risk, although mainly it's because they are seeking reassurance. Ultimately, people want to be told that things will be OK, and that their particular story will have a happy ending. (I've experimented with telling people the opposite, suggesting to audiences that there could be trouble ahead, but with the exception of risk professionals, nobody appears to listen, much less act.)

What else can I tell you? My first tip would be to think in terms of probabilities, especially high-probability/high-impact events, although not to the exclusion of low-probability/high-impact events, which are the kinds of things that are often hidden in plain sight and catch us unawares. The global COVID-19 pandemic of 2020 and 2021 and the 2008 global financial crash are prime examples.

One of the ways I personally like to think about the future is by creating diagrams. One 'map' that I created a few years ago at Imperial College London looked at emerging developments in science and technology, but made a clear distinction between the probable and the possible (the definition of possible being 'not impossible', although sometimes this is harder than you might imagine). Visual representations allow for connections to be explored, and can highlight both the cycles and counter-trends that frequently exist, creating friction or moving developments in a different direction to the original trend or driving force.

Another way of wrestling the future into mild submission is to reject its very existence since, ultimately, the only certainty is that it's uncertain. You can think about this in two ways. The first is to live in the moment and not worry about the future, adopt a stoic mindset and only worry about what you can influence or control.

The second is to make mild preparations for multiple eventualities while maintaining a healthy degree of scepticism about those who insist upon certain singular outcomes or inevitabilities.

Organisations often develop what I call official futures over time. These are singular futures, a form of Group Think, created or endorsed by the CEO, that use words which effectively close down debate about particular scenarios. 'Could', 'might', or 'should' are very important words, because they allow for further investigation.

Adopting a scenario-thinking approach that acknowledges ambiguity and the existence of multiple futures, is a productive way of dealing with these issues. It is also a good way to interrogate the hidden assumptions and biases that are contained within all forecasts and predictions.

What else have I learnt over my years as a futurist? One observation is that it's hard to lure people away from their everyday busyness to think about their future, unless they are feeling either playful/adventurous or vulnerable/scared. Vulnerable seems to work best. The threat of imminent extinction is even better. This applies to innovation, too – possibly because it eliminates superfluous distractions and allows us to focus on what's really important, not what's momentarily urgent. Good examples might include the current rise of China, artificially intelligent autonomous systems, and climate change.

Another point to highlight is our tendency to extrapolate from recent experience. People consider what's happening *now* (or focus on very recent data/experience) and project this forward, in a linear manner. For example, at the time of writing, some commentators are suggesting that COVID-19 means 'handshakes are history'. This isn't impossible, but I think it highly unlikely. Chances are it's a knee-jerk and somewhat simplistic forward projection.

If it helps, think in terms of the present as having a gravitational pull. If present conditions are dark or dystopian, the prevailing mood will be projected on to all imagined futures, almost as a mass contagion, and vice versa. The best way to avoid this is to set any discussion sufficiently into the future to achieve objectivity.

In practical terms, this means thinking at least ten and ideally fifteen years ahead, although if you travel too far things can get a little silly.

Finally, I firmly believe that while scenarios, weak signals, and wild cards ought to be carefully monitored, it is incumbent upon us all to participate in the creation of a preferred future, one that is fully inclusive and open to all, otherwise we may one day find ourselves forced into a future that is not to our liking. We ought to therefore spend less time individually worrying about bad things that might happen, and more time collectively teasing apart what a good life might look like, which is obviously the oldest philosophical question around. In other words, we ought to think of ourselves as agents of change: citizens and activists who are collectively empowered to imagine and create positive futures rather than atomised and weak individuals who can do nothing more than react and respond.

In my view, the biggest challenge the world faces at the moment is that there is no commonly held vision of where the human race is heading, except for one created by the tech sector.

Is this a future you are happy about? If not, get involved in the future and do something about it. You might strive to actively raise awareness about the ethical challenges of technology, especially concerns relating to privacy, surveillance, human agency, and environmental impact. Alternatively, you might fight for the rights of battlefield robots or robotic sex workers. It's your future: be in it.

Like sci-fi, one of the roles of futurology is to reflect and, if necessary, forewarn. Back in 1970, *Future Shock*, by Alvin Toffler, warned that the perception of too much change over too short a period of time would create mental instability.[3] Perception is personal, but in my opinion, the absence of a shared vision for humanity is making matters worse. We are not hostages to the future. We cannot predict the future, but we can shape it through discussion, and, most importantly, by the actions that you and I make each and every day – especially, in this instance, as consumers of technology. So, for me, the question is not whether sex robots are the future but, rather, whether this is the future we want.[4] During the pandemic, I was often asked what I thought would happen next. But that was always the wrong question. The question was, and remains, what do we want to happen next? What kind of future do we all want?

Pygmalion

had seen them, spending their lives in wickedness, and, offended by the failings that nature gave the female heart, he lived as a bachelor, without a wife or partner for his bed. But, with wonderful skill, he carved a figure, brilliantly, out of snow-white ivory, no mortal woman, and fell in love with his own creation.

The features are those of a real girl, who, you might think, lived, and wished to move, if modesty did not forbid it. Indeed, art hides his art.

He marvels: and passion for this bodily image consumes his heart. Often, he runs his hands over the work, tempted as to whether it is flesh or ivory, not admitting it to be ivory.

He kisses it and thinks his kisses are returned; and speaks to it; and holds it, and

imagines that his fingers press into the limbs, and is afraid lest bruises appear from the pressure. Now he addresses it with compliments, now brings it gifts that please girls, shells and polished pebbles, little birds, and many-coloured flowers, lilies and tinted beads, and the Heliades's amber tears, that drip from the trees.

He dresses the body, in clothing; places rings on the fingers; places a long necklace round its neck; pearls hang from the ears, and cinctures round the breasts. All are fitting: but it appears no less lovely, naked.

He arranges the statue on a bed on which cloths dyed with Tyrian murex are spread, and calls it his bedfellow, and rests its neck against soft down, as if it could feel.

The day of Venus's festival came, celebrated throughout Cyprus, and heifers, their curved horns gilded, fell, to the blow on their snowy neck. The incense was smoking, when Pygmalion, having made his offering, stood by the altar, and said, shyly: 'If you can grant all things, you gods,

I wish as a bride to have ...' and not daring to say, 'the girl of ivory,' he said,

'one like my ivory girl.' Golden Venus, for she herself was present at the festival, knew what the prayer meant, and as a sign of the gods' fondness for him, the flame flared three times, and shook its crown in the air. When he returned, he sought out the image of his girl, and leaning over the couch, kissed her. She felt warm: he pressed his lips to her again, and also touched her breast with his hand. The ivory yielded to his touch, and lost its hardness, altering under his fingers, as the bees' wax of Hymettus softens in the sun, and is moulded, under the thumb, into many forms, made usable by use. The lover is stupefied, and joyful, but uncertain, and afraid he is wrong, reaffirms the fulfillment of his wishes, with his hand, again, and again.

It was flesh! The pulse throbbed under his thumb. Then the hero of Paphos, was indeed overfull of words with which to thank Venus, and still pressed his mouth against a mouth that was not merely a likeness. The girl felt the kisses he gave, blushed, and, raising her bashful eyes to the light, saw both her lover and the sky. The goddess attended the marriage that she had brought about, and when the moon's horns had nine times met at the full, the woman bore a son, Paphos, from whom the island takes its name.

(*Metamorphoses*, Book X, Ovid (8 CE), trans. Anthony S. Kline, 2000)

```
/* mouth against mouth */

99: ba 30f;                                    \
   a, b, %o3;                                  \
   .section __ex_table,ALLOC;                  \
   .text;                                      \
   .align  4

   /* 12 superscalar cycles seems to be the
    * limit for this case, so we do all the
    * ldd's together to get Viking MXCC into
    * streaming mode. Ho hum... */

   ldd     [src + off + 0x10], t4;    \
   st      t0, [dst + off + 0x00];    \
   addxcc  t0, sum, sum;              \
   st      t1, [dst + off + 0x04];    \

   /* Yuck, 6 superscalar cycles...
    * Also, handle the alignment code
    * out of band. */

cc_dword_align:
   addcc %g4, %g7, %g7
      ! he carved a figure
   add    %o1, 2, %o1
      ! out of snow white ivory
   srl    %g7, 16, %g3
      ! no mortal woman
```

```
    addx   %g0, %g3, %g4
      ! the features are those of a real girl
    sll  %g7, 16, %g7
    sll  %g4, 16, %g3
      ! he runs his hand over the work

/* mouth against mouth */
/* mouth against mouth */

1: be       3f
     andcc  %g1, 0xffffff80, %g0
   EX2(stb %o5, [%o1 + 0x00])
    sub      %g1, 4, %g1
      ! he kisses it and thinks
   EX2(st  %g4, [%o1 + 0x00])
      ! his kisses returned
    add      %o0, 4, %o0
      ! fingers press into the limbs
    addcc  %g4, %g7, %g7
    add      %o1, 4, %o1
      ! no less lovely naked
    b         3f
     andcc  %g1, 0xffffff80, %g0


/* Sun, you just can't beat me, you just
 * can't.  Stop trying, give up.
 * I'm serious, I am going to kick the
 * living shit out of you, game over,
 * lights out. */
```

SWEET, ROBOT

sometimes love greets you warm

    like dry clothes hanging on a line

where ice cream melts down

    your hand and her arms

all you taste is the sweet and salt of her.

all her metal eventually in your mouth

    changing everything.

do you remember how it began?

    evening fell, and she asked, *can i take you on a walk?*

then, *can i kiss you?*

you both entered the part of the forest so deep

    there are only echoes there.

soft light.

        dust dance like stars.

dusky beautiful birds disappear,

    one by one,

from the corners of your eyes.

you fall deeply into the small of moonlight.

        fall deeply into circuits and glow.

im still learning how to listen, you confess.

    im still learning how to walk as i'm learning how to ask, she says.

but here we are.

(Margaret Rhee, *Love, Robot*, 2017)

Notes

1 Made, Not Born: The Ancient History of Intelligent Machines

1 Aifric Campbell, *Scarlett and Gurl*, this volume.

2 Homer's *Iliad*: self-pumping bellows, 18.468–73; tripod servants, 18.373–9; golden handmaids, 18.412–22.

3 Homer's *Odyssey*: 7.36; 8.555–64. See Genevieve Liveley and Samantha Thomas, 'Homer's Intelligent Machines: AI in Antiquity', in *AI Narratives: A History of Imaginative Thinking about Intelligent Machines*, ed. Stephen Cave, Kanta Dihal, and Sarah Dillon (Oxford: Oxford University Press, 2020), 25–48.

4 Adrienne Mayor, *Gods and Robots: The Ancient Quest for Artificial Life* (Princeton, NJ: Princeton University Press, 2018), 7–32.

5 Liveley and Thomas, 'Homer's Intelligent Machines', 40.

6 Sylvia Berryman, 'Ancient Automata and Mechanical Explanation', *Phronesis* 48 (2003): 344–69; Susan Murphy, 'Heron of Alexandria's *On Automaton-Making*', *History of Technology* 17 (1995): 1–45.

7 Karin Tybjerg, 'Wonder-Making and Philosophical Wonder in Hero of Alexandria', *Studies in the History and Philosophy of Science* 34 (2003): 443–66; Paul Keyser, 'Venus and Mercury in the Grand Procession of Ptolemy II', *Historia* 65 (2016): 31–52.

8 Mayor, *Gods and Robots*, 203–5; Signe Cohen, 'Romancing the Robot and Other Tales of Mechanical Beings in Indian Literature', *Acta Orientalia* 64 (2002): 65–75; Daud Ali, 'Bhoja's Mechanical Garden: Translating Wonder across the Indian Ocean, circa 800–1100 CE', *History of Religions* 55 (2016): 460–93, 481–4.

9 Joseph Needham, *Science and Civilization in China: Mechanical Engineering* (Taipei: Caves Books, 1986), 159–60; Michelle Wang, 'Early Chinese Buddhist Sculptures as Animate Bodies and Living Presences', *Ars Orientalis* 46 (2016): 13–38.

10 E. R. Truitt, ' "*Trei poëte, sages dotors, qui mout sorent di nigromance*": Knowledge and Automata in Twelfth-Century French Literature', *Configurations* 12 (2004): 167–93.

11 Donald Hill, 'Medieval Arabic Mechanical Technology', in *Studies in Medieval Islamic Technology: From Philo to al-Jazari – from Alexandria to Diyar Bakr*, ed. David A. King (Aldershot: Ashgate Variorum, 1998), 222–37; Al-Jazari, *Kitab al-Jami'bayn al'ilm wa-'l-'amal al-nafi' fi sinat'at al-hiyal*, in English as *The Book of Ingenious Mechanical Devices*, trans. Donald R. Hill (Dordecht: D. Reidel, 1974). Automata also appear in the literature of the Islamicate world, often in the context of the rulers of ancient Egypt.

12 Jan M. Ziolkowski and Michael J. Putnam, eds., *The Virgilian Tradition: The First Fifteen Hundred Years* (New Haven, CT: Yale University Press, 2008), 856; E. R. Truitt, 'Demons and Devices: Artificial and Augmented Intelligence before AI', in *AI Narratives: A History of Imaginative Thinking about Intelligent Machines*, ed. Stephen Cave, Kanta Dihal, and Sarah Dillon (Oxford: Oxford University Press, 2020), 49–71.

13 Penny Sullivan, 'Medieval Automata: The "Chambre de Beautés" in Benoît de Sainte-Maure's *Roman de Troie*', *Romance Studies* 6 (1985): 1–20; Truitt, ' "*Trei poëte*".

14 E. R. Truitt, *Medieval Robots* (Philadelphia: University of Pennsylvania Press, 2015), 130–5.

15 Truitt, *Medieval Robots*, 132.

16 Truitt, 'Demons and Devices', 53–4.

17 Thomas of Britain, *The Romance of Tristram and Ysolt by Thomas of Britain*, trans. Roger Loomis (New York: Columbia University Press, 1931); Paul Schach, ed. and trans., *The Saga of Tristram and Ísönd* (Lincoln: University of Nebraska Press, 1973), discussed in Truitt, *Medieval Robots*, 100–2.

18 These texts were compiled between the second century BCE and the fourth century CE. Cohen, 'Romancing the Robot'; V. Raghavan, *Yantras or Mechanial Contrivances in Ancient India* (Bangalore: Indian Institute of Culture, 1952). On the more recent history of sex robots, see Kate Devlin, *Turned On: Science, Sex, and Robots* (London: Bloomsbury, 2018).

19 V. Raghavan, 'Somadeva and King Bhoja', *Journal of the University of Gauhati* 3 (1952): 35–8; Ali, 'Bhoja's Mechanical Garden'.

20 Nancy Siraisi, *Medieval and Early Renaissance Medicine* (Chicago: University of Chicago Press, 1990), 101–8.

21 John Cohen, *Human Robots in Myth and Science* (London: Allen and Unwin, 1966).

22 *The Book of Liezi* purports to be from the fourth century BCE but may in fact date from much later, around 300 CE. Mayor, *Gods and Robots*, 121. On later legends, see Truitt, *Medieval Robots*, 91–4; Minsoo Kang, 'The Mechanical Daughter of René Descartes: The Origin and History of an Intellectual Fable', *Modern Intellectual History* 14 (2016): 633–60.

23 E. T. A. Hoffman, 'The Sandman', in *Tales of Hoffmann*, ed. and trans. R. J. Hollingdale (London: Penguin, 1982), 85–126.

24 For example, Ruha Benjamin, ed., *Captivating Technology: Race, Carceral Technoscience, and Liberatory Imagination in Everyday Life* (Durham, NC: Duke University Press, 2019); Michal Luria, Juliet Pusateri, Judeth Oden Choi, Reuben Aronson, Nur Yildrim, and Molly Wright Steenson, 'Medieval Robots: The Role of Historical Automata in the Design of Future Robots', *Companion Publication of the 2020 ACM Designing Interactive Systems Conference* (July 2020): 191–5, https://doi.org/10.1145/3393914.3395890.

2 Roxanne, or Companion Robots and Hierarchies of the Human

1 Aifric Campbell, *Scarlett and Gurl*, this volume.

2 Joanna Bourke, *What It Means to Be Human: Reflections from 1791 to the Present* (London: Virago, 2011).

3 Simone de Beauvoir, *The Second Sex*, trans. Constance Borde and Sheila Malovany-Chevallier (New York: Vintage Books, 2011), 26.

4 Alexander G. Weheliye, *Habeas Viscus: Racializing Assemblages, Biopolitics, and Black Feminist Theories of the Human* (Durham, NC: Duke University Press, 2014), 3.

5 Campbell, *Scarlett and Gurl*, this volume.

6 Ibid., p. 114.

7 Ibid., p. 109.

8 Julia O'Connell Davidson, *Prostitution, Power and Freedom* (New York: John Wiley & Sons, 2013).

9 Sinziana Gutiu, *Sex Robots and Roboticization of Consent* (Ottawa: We Robot, University of Ottawa, 2012), http://robots.law.miami.edu/sinziana-gutiu-on-sex-robots-and-roboticization-of-consent/.

10 Campbell, *Scarlett and Gurl*, this volume, p. 116.

11 'Immanuel Kant's Lectures on Ethics', in *Lectures on Ethics*, ed. Peter Heath and J. B. Schneewind (Cambridge, UK: Cambridge University Press, 1997), 212, https://doi.org/10.1017/CBO9781107049512.

12 Joanna Stern, 'Alexa, Siri, Cortana: The Problem with All-Female Digital Assistants', *Wall Street Journal* (sec. Tech.), (21 February 2017), www.wsj.com/articles/alexa-siri-cortana-the-problem-with-all-female-digital-assistants-1487709068.

13 Nóra Ni Loideain and Rachel Adams, 'From Alexa to Siri and the GDPR: The Gendering of Virtual Personal Assistants and the Role of Data Protection Impact Assessments', *Computer Law & Security Review* 36, no. 105366 (9 December 2019), https://doi.org/10.1016/j.clsr.2019.105366.

14 Campbell, *Scarlett and Gurl*, this volume, p. 210.

15 Ibid., p. 42.

16 Ibid., p. 74.

17 Ibid., p. 38.

18 Ibid., p. 115.

19 Ibid., p. 209.

20 Ibid., p. 121.

21 Ibid., p. 118.

22 Robertson, Jennifer Ellen, *Robo Sapiens Japanicus* (Oakland: University of California Press, 2017).

23 Stephen Cave and Kanta Dihal, 'The Whiteness of AI', *Philosophy & Technology* 33 (6 August 2020): 685–703, https://doi.org/10.1007/s13347-020-00415-6.

24 E. M. Forster, 'The Machine Stops', *Oxford and Cambridge Review* (November 1909), http://archive.ncsa.illinois.edu/prajlich/forster.html; Isaac Asimov, *The Robots of Dawn* (New York: Bantam Books, 1983); Andrew Stanton, *WALL-E* (Burbank, CA: Disney, 2008), www.imdb.com/title/tt0910970/.

25 Campbell, *Scarlett and Gurl*, this volume, p. 115.

26 Sherry Turkle, *Alone Together: Why We Expect More from Technology and Less from Each Other* (New York: Basic Books, 2011).

27 Campbell, *Scarlett and Gurl*, this volume, p. 158.

3 while (alive) {love me;}

1 Nicola Döring and Sandra Poeschl, 'Love and Sex with Robots: A Content Analysis of Media Representations', *International Journal of Social Robotics* 11, no. 4 (2019): 665–77.

2 Kate Devlin, *Turned On: Science, Sex and Robots* (London: Bloomsbury, 2018).

3 Despina Kakoudaki, *Anatomy of a Robot: Literature, Cinema, and the Cultural Work of Artificial People* (New Brunswick, NJ: Rutgers University Press, 2014); Genevieve Liveley and Samantha Thomas, 'Homer's Intelligent Machines: AI in Antiquity', in *AI Narratives: A History of Imaginative Thinking about Intelligent Machines*, ed. Stephen Cave, Kanta Dihal, and Sarah Dillon (Oxford: Oxford University Press, 2020), 25–48.

4 Sven Behnke, 'Humanoid Robots: From Fiction to Reality?' *KI 22* 4 (2008): 5–9; Joe Denny, Mohamed Elyas, Shannon Angel D'costa, and Royson Donate D'Souza, 'Humanoid Robots: Past, Present and the Future', *European Journal of Advances in Engineering and Technology* 3, no. 5 (2016): 10 (see also 8–15).

5 Masahiro Mori, '*Bukimi no tani* (The Uncanny Valley)', *Energy* 7 (1970): 33–5, trans. Karl F. MacDorman and Norri Kageki, *IEEE Spectrum* (12 June 2012), https://spectrum.ieee.org/automaton/robotics/humanoids/the-uncanny-valley.

6 Karl F. MacDorman, 'Mortality Salience and the Uncanny Valley', *5th IEEE-RAS International Conference on Humanoid Robots* (2005): 399–405.

7 Jari Kätsyri, Klaus Förger, Meeri Mäkäräinen, and Tapio Takala, 'A Review of Empirical Evidence on Different Uncanny Valley

Hypotheses: Support for Perceptual Mismatch as One Road to the Valley of Eeriness,' *Frontiers in Psychology* 6 (2015): 390; Maya B. Mathur and David B. Reichling, 'Navigating a Social World with Robot Partners: A Quantitative Cartography of the Uncanny Valley,' *Cognition* 146 (2016): 22–32.

8 Angela Tinwell, Mark Grimshaw, and Andrew Williams, 'The Uncanny Wall,' *International Journal of Arts and Technology* 4, no. 3 (2011): 326–41.

9 Clifford Nass, Jonathan Steuer, and Ellen R. Tauber, 'Computers Are Social Actors,' *Proceedings of the SIGCHI conference on Human Factors in Computing Systems* (1994): 72–8; Matthias Hoenen, Katrin T. Lübke, and Bettina M. Pause, 'Non-Anthropomorphic Robots as Social Entities on a Neurophysiological Level,' *Computers in Human Behavior* 57 (2016): 182–6.

10 Ewart J. de Visser, Samuel S. Monfort, Ryan McKendrick, Melissa A. B. Smith, Patrick E. McKnight, Frank Krueger, and Raja Parasuraman, 'Almost Human: Anthropomorphism Increases Trust Resilience in Cognitive Agents,' *Journal of Experimental Psychology: Applied* 22, no. 3 (2016): 331.

11 Julie Wosk, *My Fair Ladies: Female Robots, Androids, and Other Artificial Eves* (New Brunswick, NJ: Rutgers University Press, 2015).

12 Judy Wajcman, *Feminism Confronts Technology* (Cambridge, UK: Polity, 1991), 22–3; Jennifer Robertson, 'Gendering Humanoid Robots: Robo-Sexism in Japan,' *Body & Society* 16, no. 2 (2010): 2 (see also 1–36).

13 Kate Devlin and Chloé Locatelli, 'Guys and Dolls: Sex Robot Creators and Consumers,' in *Maschinenliebe* ed. Oliver Bendel (Heidelberg: Springer Gabler, 2020), 81 (see also 79–92).

14 Sarah Hatheway Valverde, 'The Modern Sex Doll-Owner: A Descriptive Analysis' (Master's thesis, California Polytechnic State University, 2012); Norman Makoto Su, Amanda Lazar, Jeffrey Bardzell, and Shaowen Bardzell, 'Of Dolls and Men: Anticipating Sexual Intimacy with Robots,' *ACM Transactions on Computer-Human Interaction (TOCHI)* 26, no. 3 (2019): 1–35.

15 Danielle   Knafo   and   Rocco   Lo   Bosco,   *The   Age   of Perversion: Desire and Technology in Psychoanalysis and Culture* (London: Routledge, 2016).

16 Kate Devlin and Olivia Belton, 'The Measure of a Woman: Fembots, Fact and Fiction', in *AI Narratives: A History of Imaginative Thinking about Intelligent Machines*, ed. Stephen Cave, Kanta Dihal, and Sarah Dillon (Oxford: Oxford University Press, 2020), 359 (see also 358–82).

17 Allison P. Davis, 'Are We Ready for Robot Sex?' *New York Magazine* (14–27 May 2018), https://nymag.com/press/2018/05/on-the-cover-are-we-ready-for-robot-sex.html.

18 Yolande Strengers and Jenny Kennedy, *The Smart Wife: Why Siri, Alexa, and Other Smart Home Devices Need a Feminist Reboot* (Cambridge, MA: MIT Press, 2020).

19 Alisha Pradhan, Leah Findlater, and Amanda Lazar, ' "Phantom Friend" or "Just a Box with Information": Personification and Ontological Categorization of Smart Speaker-based Voice Assistants by Older Adults', *Proceedings of the ACM on Human–Computer Interaction* 3, no. CSCW (2019): 1–21.

20 Joanna Rothkopf, 'Single? Consider a Cross-Dimensional (Human-Hologram) Marriage', *Esquire* (13 November 2018), www.esquire.com/lifestyle/a25018920/japan-married-hologram-gatebox/.

21 Patricia J. Bota, Chen Wang, Ana L. N. Fred, and Hugo Plácido Da Silva, 'A Review, Current Challenges, and Future Possibilities on Emotion Recognition Using Machine Learning and Physiological Signals', *IEEE Access* 7 (2019): 140990–1020.

22 Melanie Birks, Marie Bodak, Joanna Barlas, June Harwood, and Mary Pether, 'Robotic Seals as Therapeutic Tools in an Aged Care Facility: A Qualitative Study', *Journal of Aging Research* 2016, no. 2 (2016): 1–7, https://doi.org/10.1155/2016/8569602; Kathleen Kara Fitzpatrick, Alison Darcy, and Molly Vierhile, 'Delivering Cognitive Behavior Therapy to Young Adults with Symptoms of Depression and Anxiety Using a Fully Automated Conversational Agent (Woebot): A Randomized Controlled Trial', *JMIR Mental Health* 4, no. 2 (2017): e19.

23 See https://futureofsex.net/sex-tech/.

24 Marcus Dowling, 'Growth of Sexual Wellness Market Speeding Up and Will Surpass $123B by 2026, Says Lovense CEO', *Future of Sex* (4 September 2020), https://futureofsex.net/sex-tech/growth-of-sex-tech-market-speeding-up-and-will-surpass-123b-by-2026-says-lovense-ceo/.

25 Hayley   Campbell,   'Better   Loving   through   Technology: A Day at the Sex-Toy Hackathon', *Observer* (10 December 2017), www.theguardian.com/technology/2017/dec/10/better-loving-through-technology-sex-toy-hackathon.

26 Vaughan Bell, 'Don't Touch That Dial! A History of Media Technology Scares, from the Printing Press to Facebook', *Slate* (15   February   2010),   https://slate.com/technology/2010/02/a-history-of-media-technology-scares-from-the-printing-press-to-facebook.html;   Adam   Jezard,   'Technophobia   Is so Last Century: Fears of Robots, AI and Drones Are Not New', *Financial Times* (2 March 2016), www.ft.com/content/a9ec6360-cf80-11e5-92a1-c5e23ef99c77.

4 Robots as Solace and the Valence of Loneliness

1 Cynthia Breazeal, Jesse Gray, Guy Hoffman, and M. Berlin, 'Social Robots: Beyond Tools to Partners', *RO-MAN 2004, 13th IEEE International Workshop on Robot and Human Interactive Communication (IEEE Catalog No. 04TH8759)* (2004): 551–6; Julie Carpenter, *Culture and Human–Robot Interaction in Militarized Spaces: A War Story* (London: Routledge, 2016); Leila Takayama, Wendy Ju, and Clifford Nass, 'Beyond Dirty, Dangerous and Dull: What Everyday People Think Robots Should Do', *2008 3rd ACM/IEEE International Conference on Human-Robot Interaction (HRI)* (2008): 25–32.

2 Carpenter, *Culture and Human–Robot Interaction.*

3 Ibid., 106.

4 Ibid.

5 Ibid., 117; K. Hay, personal communication, 2 October 2013.

6 Brenda Leong and Evan Selinger, 'Robot Eyes Wide Shut: Understanding Dishonest Anthropomorphism', *Proceedings of the Conference on Fairness, Accountability, and Transparency* (January 2019): 299–308, https://doi.org/10.1145/3287560.3287591; Kathleen Richardson, 'Sex Robot Matters: Slavery, the Prostituted, and the Rights of Machines', *IEEE Technology and Society Magazine* 35, no. 2 (2016): 46–53; John Danaher, 'Robot Betrayal: A Guide to the Ethics of Robotic Deception', *Ethics and Information Technology* (2020): 1–12; Sherry Turkle, *Alone Together: Why We Expect More from Technology and Less from Each Other* (London: Hachette, 2017).

7 Richardson, 'Sex Robot Matters'; Turkle, *Alone Together*.

8 Peter H. Kahn, Aimee L. Reichert, Heather E. Gary, Takayuki Kanda, Hiroshi Ishiguro, Solace Shen, Jolina H. Ruckert, and Brian Gill, 'The New Ontological Category Hypothesis in Human–Robot Interaction', *6th ACM/IEEE International Conference on Human–Robot Interaction (HRI)* (2011): 159–60; Carpenter, *Culture and Human–Robot Interaction*.

9 John T. Cacioppo and Stephanie Cacioppo, 'The Growing Problem of Loneliness', *The Lancet* 391, no. 10119 (2018): 426. See also Kerstin Dautenhahn, Sarah Woods, Christina Kaouri, Michael L. Walters, Kheng Lee Koay, and Iain Werry, 'What Is a Robot Companion: Friend, Assistant or Butler?' *IEEE/RSJ International Conference on Intelligent Robots and Systems* (2005): 1192–7; Victoria Groom, Leila Takayama, Paloma Ochi, and Clifford Nass, 'I Am My Robot: The Impact of Robot-Building and Robot Form on Operators', *2009 4th ACM/IEEE International Conference on Human-Robot Interaction (HRI)* (2009): 31–6; Kathleen Richardson, *An Anthropology of Robots and AI: Annihilation Anxiety and Machines* (London: Routledge, 2015).

5 Robot Nannies Will Not Love

1 Erasmus Darwin, 'Of Instinct', in *Zoonomia; or, The Laws of Organic Life*, vol. 1 (3rd American ed., corr.) (Boston, MA: Thomas & Andrews, 1809), 101–42.

2 Jodi Forlizzi and Carl DiSalvo, 'Service Robots in the Domestic Environment', *Proceedings of the 1st ACM SIGCHI/SIGART Conference on Human–Robot Interaction* (March 2006): 258–65.

3 S. M. Shuster, E. V. Lonsdorf, G. M. Wimp, J. K. Baily, and T. G. Whitham, 'Community Heritability Measures the Evolutionary Consequences of Indirect Genetic Effects on Community', *Structure Evolution* 60, no. 5 (2007): 991–1003.

4 Michael A. Arbib and Jean Marc Fellous, 'Emotions: From Brain to Robot', *Trends in Cognitive Sciences* 8, no. 12 (1 December 2004): 554–61.

5 Joanna J. Bryson, 'Robots Should Be Slaves', *Close Engagements with Artificial Companions* (Amsterdam: John Benjamins, 2010), 63–74; Joanna J. Bryson, Mihailis E. Diamantis, and Thomas D. Grant, 'Of, For, and By the People: The Legal Lacuna of Synthetic Persons', *Artificial Intelligence and Law* 3, no. 25 (2017): 273–91; Organisation for Economic Co-Operation and Development (OECD), 'Recommendation of the Council on Artificial Intelligence', in *OECD Legal Instruments OECD/LEGAL/0449* (Paris: Organisation for Economic Cooperation and Development, May 2019).

6 S. S. Walsh, S. N. Jarvis, E. M. Towner, and A. Aynsley-Green, 'Annual Incidence of Unintentional Injury among 54,000 Children', *Injury Prevention* 2, no. 1 (1996): 16–20; Rosana E. Norman, Munkhtsetseg Byambaa, Rumma De, Alexander Butchart, James Scott, and Theo Vos, 'The Long-Term Health Consequences of Child Physical Abuse, Emotional Abuse, and Neglect: A Systematic Review and Meta-Analysis', *PLOS Medicine* 9, no. 11: e1001349 (27 November 2012), https://doi.org/10.1371/journal.pmed.1001349.

7 Patrick Lavelle, 'Product Liability', in *Forensic Science and Law: Investigative Applications in Criminal, Civil and Family Justice*, ed. Cyril H. Wecht and John T. Rago (Boca Raton: CRC Press, 2005), 231–8.

8 D. DiLillo, A. Damashek, and L. Peterson, 'Maternal Use of Baby Walkers with Young Children: Recent Trends and Possible Alternatives', *Injury Prevention* 7, no. 3 (2001): 223–7; J. S. Tay and J. S. Garland, 'Serious Head Injuries from Lawn Darts', *Pediatrics* 79 (1987): 261–3; Sara V. Sotiropoulos, Mary Anne Jackson, Gerald F. Tremblay, Fred Burry, and Lloyd C. Olson, 'Childhood Lawn Dart

Injuries: Summary of 75 Patients and Patient Report', *American Journal of Diseases of Children* 144, no. 9 (1990): 980–2.

9 Peter Newman and Jeff Kenworthy, 'Costs of Automobile Dependence: Global Survey of Cities', *Transportation Research Record* 1670 (1999): 17–26.

10 Timothy D. Lytton, 'Using Litigation to Make Public Health Policy: Theoretical and Empirical Challenges in Assessing Product Liability, Tobacco, and Gun Litigation', *Journal of Law, Medicine and Ethics* 32, no. 4 (2019): 556–64; Eliana Rae Eitches, 'The Protection of Lawful Commerce in Arms Act: Precipitating the World's Most Dangerous Game', *SSRN Electronic Journal* (10 May 2017), https://dx.doi.org/10.2139/ssrn.3009413.

11 Benjamin C. Zipursky, 'Rights, Wrongs, and Recourse in the Law of Torts', *Vanderbilt Law Review* 51 (1998), https://ir.lawnet.fordham.edu/faculty_scholarship/840.

12 Robert W. Kolb, 'Ford Pinto', in *The SAGE Encyclopedia of Business Ethics and Society*, ed. Robert W. Kolb (London: Sage, 2018).

13 Benjamin C. Zipursky, 'Palsgraf, Punitive Damages, and Pre-emption', *Harvard Law Review* 125, no. 1757 (18 July 2012).

14 John M. Purvis and Stuart A. Hirsch, 'Playground Injury Prevention', *Clinical Orthopaedics and Related Research (1976–2007)* 409 (2003): 11–19, https://doi.org/10.1097/01.blo.0000057780.39965.2c.

15 Aifric Campbell, *Scarlett and Gurl*, this volume, p. 37.

16 Alexander Alspach, Joohyung Kim, and Katsu Yamane, 'Design of a Soft Upper Body Robot for Physical Human–Robot Interaction', *IEEE–RAS International Conference on Humanoid Robots* (2015): 290–6; Si Yu Sun, Wei Wei Xu, Zhuo Hang Li, Kwan-Keung Ng, and Ivan Ka-Wai Lai, 'A Study on the Appearances and Functionalities of Education Robots for Attracting Students' Attention and Interactive Interests', *Proceedings – 2018 International Symposium on Educational Technology* (2018): 245–9.

17 Joanna J. Bryson and Emmanuel Tanguy, 'Simplifying the Design of Human-Like Behaviour: Emotions as Durative Dynamic State for Action Selection, Creating Synthetic Emotions through Technological and Robotic Advancements', *IGI Global* (2012): 32–53.

18 Stéphanie Bioulac, Lisa Arfi, and Manuel P. Bouvard, 'Attention Deficit/Hyperactivity Disorder and Video Games: A Comparative Study of Hyperactive and Control Children', *European Psychiatry* 23, no. 2 (2008): 134–41, https://doi.org/10.1016/j.eurpsy.2007.11.002.

19 Laura Canetti, Eytan Bachar, E. Galili-Weisstub, and B. Kaplan De-Nour, 'Parental Bonding and Mental Health in Adolescence', *Adolescence* 32 (1997): 381–94.

20 David Weibel, Bartholomäus Wissmath, Stephan Habegger, and Yves Steiner, 'Playing Online Games against Computer vs. Human-Controlled Opponents: Effects on Presence, Flow, and Enjoyment', *Computers in Human Behavior* 24, no. 5 (2008): 2274–91.

21 Elizabeth A. Vandewater, Victoria J. Rideout, Ellen A. Wartella, Xuan Huang, June H. Lee, and Misuk Shim, 'Digital Childhood: Electronic Media and Technology Use Among Infants, Toddlers, and Pre-Schoolers', *Pediatrics* 119, no. 5 (2007): e1006–15.

22 Joanna J. Bryson and Phil Kime, 'Just Another Artifact: Ethics and the Empirical Experience of AI', *Fifteenth International Congress on Cybernetics* (1998): 385–90.

23 United Nations, *A Summary of the UN Convention on the Rights of the Child* (London: UNICEF, 1992).

24 Lundy Laura, ' "Voice" Is Not Enough: Conceptualising Article 12 of the United Nations Convention on the Rights of the Child', *British Educational Research Journal* 33, no. 6 (2007): 927–42.

25 Gerison Lansdown, *Every Child's Right to Be Heard* (London: Save the Children, 2011).

26 Alex Sciuto, Arnita Saini, Jodi Forlizzi, and Jason I. Hong, ' "Hey Alexa, What's Up?": Studies of In-Home Conversational Agent Usage, *DIS* 2018', *Proceedings of the 2018 Designing Interactive Systems Conference* (2018): 857–68, https://doi.org/10.1145/3196709.3196772.

6 Masters and Servants: The Need for Humanities in an AI-Dominated Future

1 Alan M. Turing, AMT/C 22: Unpublished manuscripts and drafts, 'The Rules of GO (A Game For Two Players with Counters)'. Written

on King's College, Cambridge writing paper, one sheet in envelope (undated), http://turingarchive.org/browse.php/C/22.

2 Alan M. Turing, 'Computing Machinery and Intelligence', *Mind* LIX, no. 236 (1950): 433–60.

3 Yonhap News Agency, 'Interview with Lee Sedol: Go Master Lee Says He Quits Unable to Win Over AI Go Players' (27 November 2019), https://en.yna.co.kr/view/AEN20191127004800315. See also Greg Kohs (dir.), *AlphaGo: The Movie* (Moxie Pictures, 2017), www.youtube.com/watch?v=WXuK6gekU1Y; and David Silver, Julian Schrittwieser, Ioannis Antonoglu, Aja Huang, Arthur Guez, Thomas Hubert, Lucas Baker, Matthew Lai, Adrian Bolton et al., 'Mastering the Game of Go Without Human Knowledge', *Nature* 550 (2017), 354–9, https://doi.org/10.1038/nature24270.

4 Tom Simonite, 'Who's Listening When You Talk to Your Google Assistant?' *Wired* (7 October 2019), www.wired.com/story/whos-listening-talk-google-assistant/.

5 Jessa Lingel and Kate Crawford, 'Notes from the Desk Set: A Secretary's History of Work and Surveillance', *Catalyst: Feminism, Theory, Technoscience* 6, no. 1 (2020), http://dx.doi.org/10.28968/cftt.v6i1.29949.

6 Kate Crawford and Vladan Joler, 'Anatomy of an AI System: The Amazon Echo as an Anatomical Map of Human Labor, Data and Planetary Resources', *AI Now Institute and Share Lab* (7 September 2018), https://anatomyof.ai.

7 Henry A. Kissinger, 'How the Enlightenment Ends', *The Atlantic* (June 2018), www.theatlantic.com/magazine/archive/2018/06/henry-kissinger-ai-could-mean-the-end-of-human-history/559124/; Ross Andersen, 'The Panopticon Is Already Here', *The Atlantic* (September 2020), www.theatlantic.com/magazine/archive/2020/09/china-ai-surveillance/614197/.

8 Luciano Floridi and Josh Cowls, 'A Unified Framework of Five Principles for AI in Society', *Harvard Data Science Review* 1, no. 1 (2019), https://doi.org/10.1162/99608f92.8cd550d1.

9 Charles P. Snow and Stefan Collini, 'The Rede Lecture (1959)', in *The Two Cultures* (Cambridge, UK: Canto, Cambridge University Press, [1959] 1993), 1–52.

10 Fenwick & West LLP, 'A Comparison of Large Public Companies and Silicon Valley Companies' (2018), www.fenwick.com/publication-request-gender-diversity-survey-2018; Sara A. O'Brine, 'Silicon Valley Can No Longer "Tinker Around the Edges" to Fix Its Diversity Problem', *CNN Business* (24 June 2020), https://edition.cnn.com/2020/06/24/tech/diversity-silicon-valley/index.html.

11 Meredith Broussard, *Artificial Unintelligence. How Computers Misunderstand the World* (Cambridge, MA: MIT Press, 2018).

12 Joy Buolamwini and Timnit Gebru, 'Gender Shades: Intersectional Accuracy Disparities in Commercial Gender Classification', *Proceedings of Machine Learning Research Conference on Fairness, Accountability* 81 (2018): 1–15.

13 Julia Angwin, Jeff Larson, Surya Mattu, and Lauren Kirchner, 'Machine Bias', *ProPublica* (23 May 2016), www.propublica.org/article/machine-bias-risk-assessments-in-criminal-sentencing.

14 Virginia Eubanks, *Automating Inequality: How High-Tech Tools Profile, Police, and Punish the Poor* (New York: St Martin's Press, 2018); Alexander Campolo and Kate Crawford, 'Enchanted Determinism: Power without Responsibility', *Artificial Intelligence Engaging Science Technology and Society* 6 (January 2020): 1. See also Leslie Miley, 'Thoughts on Diversity, Part 2: Why Diversity Is Difficult' (3 November 2015), https://medium.com/tech-diversity-files/thought-on-diversity-part-2-why-diversity-is-difficult-3dfd552fa1f7.

15 Touchstone Weekend leaflet, Imperial College Archives and Corporate Records Unit, GLT/1/1 (early 1950s).

16 Centre for Languages, Culture and Communication (CLCC), *Imperial Horizons: Vision 2020–2025* (January 2020), www.imperial.ac.uk/media/imperial-college/administration-and-support-services/centre-for-languages-culture-and-communications/horizons/public/Imperial-Horizons-vision-2020-2025.pdf.

17 For example, Ray Kurzweil, *The Singularity Is Near: When Humans Transcend Biology* (New York: Viking Press, 2005); Max Tegmark, *Life 3.0: Being Human in the Age of Artificial Intelligence* (New York: Knopf, 2017).

18 'Lecture to the London Mathematical Society, February 1947', in *A. M. Turing's ACE Report of 1946 and Other Papers*, ed. B. E.

Carpenter and R. W. Doran (Cambridge, MA: MIT Press, 1986), 106–24.

19 Jack B. Copeland, *Turing: Pioneer of the Information Age* (Oxford: Oxford University Press, 2013). See also Jack B. Copeland, and Diane Proudfoot, 'The Computer, Artificial Intelligence, and the Turing Test', in *Alan Turing: Life and Legacy of a Great Thinker*, ed. Christof Teuscher (Berlin: Springer, 2005), 317–52; *Daily Herald*, front page (unattributed contribution), (18 February 1947); Andrew Hodges, *Alan Turing: The Enigma, the Centenary Edition* (Princeton, NJ: Princeton University Press, 2012); and Douglas Hofstadter, and Christof Teuscher, eds., *Alan Turing* (Berlin: Springer, 2005).

7 A Feminist Artificial Intelligence?

1 Mary Shelley, *Frankenstein, or, the Modern Prometheus* (London: Lackington, Hughes, Harding, Mavor & Jones, 1818), 99.

2 Casey Chin, 'AI Is the Future – But Where Are the Women?' *Wired* (17 August 2018), www.wired.com/story/artificial-intelligence-researchers-gender-imbalance/.

3 Alan M. Turing, 'Computing Machinery and Intelligence', *Mind* LIX, no. 236 (1950): 433–60.

4 Batya Friedman and Helen Nissenbaum, 'Bias in Computer Systems', *ACM Transactions on Information Systems* 14, no. 3 (1996): 330–47; Batya Friedman and Peter H. Kahn Jr., 'Human Values, Ethics and Design', in *Handbook of Human–Computer Interaction: Fundamentals, Evolving Technologies, and Emerging Applications*, ed. Julie A. Jacko and Andrew Sears (Mahwah, NJ: Lawrence Erlbaum, 2003), 1177–201; Mary Flanagan, Daniel C. Howe, and Helen Nissenbaum, 'Embodying Values in Technology: Theory and Practice', in *Information Technology and Moral Philosophy*, ed. Jeroen van den Hoven and John Weckert (Cambridge, UK: Cambridge University Press, 2008), 322–53.

5 Bruno Latour, 'Technology Is Society Made Durable', *Sociological Review* 38, no. 1 (1990): 103–31.

6 Langdon Winner, *Autonomous Technology: Technics-Out-of-Control as a Theme in Political Thought* (Cambridge, MA: MIT Press, 1977).

7 Helen Nissenbaum, *Privacy in Context: Technology, Policy, and the Integrity of Social Life* (Palo Alto, CA: Stanford University Press, 2010).

8 Safiya Umoja Noble, *Algorithms of Oppression: How Search Engines Reinforce Racism* (New York: NYU Press, 2018).

9 Alison Adam, *Artificial Knowing: Gender and the Thinking Machine* (London: Routledge, 1998).

10 Lucy Suchman, 'Do Categories Have Politics? The Language/Action Perspective Reconsidered', *Computer Supported Cooperative Work (CSCW)* 2, no. 3 (1994): 177–90.

11 Jieyu Zhao, Tianlu Wang, Mark Yatskar, Vincente Ordonez, and Kai-Wei Chang, 'Men Also Like Shopping: Reducing Gender Bias Amplification Using Corpus-level Constraints', *Proceedings of the 2017 Conference on Empirical Methods in Natural Language Processing* (September 2017): 2979–89.

12 Sandra Harding, 'The Science Question in Feminism', *Bulletin of Science, Technology & Society* 6, no. 4 (August 1986): 400, https://doi.org/10.1177/027046768600600481.

13 Daniel Kahneman, Paul Slovic, and Amos Tversky, *Judgment Under Uncertainty: Heuristics and Biases* (Cambridge, UK: Cambridge University Press, 1982).

14 Ann Cairns, 'AI Is Failing the Next Generation of Women', *World Economic Forum* (18 January 2019), www.weforum.org/agenda/2019/01/ai-artificial-intelligence-failing-next-generation-women-bias/.

15 Kimberlé Crenshaw, 'Demarginalizing the Intersection of Race and Sex: A Black Feminist Critique of Antidiscrimination Doctrine, Feminist Theory, and Antiracist Politics', *University of Chicago Legal Forum* 1989, no. 1, art. 8 (1989): 139–67, http://chicagounbound.uchicago.edu/uclf/vol1989/iss1/8.

16 Chin, 'AI is the Future'.

17 Donald MacKenzie and Judy Wajcman, *Introduction to The Social Shaping of Technology* (Buckingham, UK: Open University Press, 1999), 3–27.

8 Colouring Outside the Lines: Constructing Racial Identity in Intelligent Machines

1 Spike Jonze, *Her* (Annapurna Pictures, 2013).

2 Christoph Bartneck, Kumar Yogeeswaran, Qi Min Ser, Graeme Woodward, Robert Sparrow, Siheng Wang, and Friederike Eyssel, 'Robots and Racism', *Proceedings of the 2018 ACM/ IEEE International Conference on Human–Robot Interaction* (2018): 196–204.

3 Stephen Cave and Kanta Dihal, 'The Whiteness of AI', *Philosophy & Technology* 33 (6 August 2020): 685–703, https://doi.org/10.1007/ s13347-020-00415-6.

4 Toby Ganley, 'What's All This Talk About Whiteness?' *Dialogue*, 1, no. 2 (2003): 12–30.

5 Laurel D. Riek and Don Howard, 'A Code of Ethics for the Human– Robot Interaction Profession', *Proceedings of We Robot* (4 April 2014), https://ssrn.com/abstract=2757805.

6 Alexander G. Weheliye, *Habeas Viscus: Racializing Assemblages, Biopolitics, and Black Feminist Theories of the Human* (Durham, NC: Duke University Press, 2014).

7 Sarah Myers West, Meredith Whittaker, and Kate Crawford, *Discriminating Systems: Gender, Race, and Power in AI* (New York: AI Now Institute, New York University, 2019).

8 Nancy Luong, 'The Race-Neutral Workplace of the Future, University of California', *UC Davis Law Review* 51, no. 719 (2017).

9 Ruha Benjamin, *Race After Technology: Abolitionist Tools for the New Jim Code* (Medford, MA: Polity, 2019), 28.

10 Yuting Liao and Jiangen He, 'The Racial Mirroring Effects on Human–Agent in Psychotherapeutic Conversation', *Proceedings of the 25th International Conference on Intelligent User Interfaces* (2020): 430–42, https://doi.org/10.1145/3377325.3377488.

11 Megan Strait, Ana Sánchez Ramos, Virginia Contreras, and Noemi Garcia, 'Robots Racialized in the Likeness of Marginalized Social Identities Are Subject to Greater Dehumanization Than Those Racialized as White', *The 27th IEEE International Symposium on Robot and Human Interactive Communication (RO-MAN)* (2018): 452–7.

12 Deborah Harrison, 'Cortana – RE-WORK Virtual Assistant Summit #reworkVA', YouTube (2016), www.youtube.com/watch?v=-WcC9PNMuL0.

13 UNESCO, EQUALS Skills Coalition, 'I'd Blush If I Could: Closing Gender Divides in Digital Skills Through Education' (2019), https://en.unesco.org/Id-blush-if-I-could.

14 Emily Lever, 'I was a Human Siri', *Intelligencer: New York Magazine* (26 April 2018), https://nymag.com/intelligencer/smarthome/i-was-a-human-siri-french-virtual-assistant.html.

15 Calvin K. Lai and Mahzarin R. Banaji, 'The Psychology of Implicit Intergroup Bias and the Prospect of Change', in *Difference without Domination: Pursuing Justice in Diverse Democracies*, ed. D. Allen and R. Somanathan (Chicago: University of Chicago Press, 2019), 115–46.

16 Samantha, © Synthea Amatus (2018).

17 Jenny Kleeman, *Sex Robots & Vegan Meat: Adventures at the Frontier of Birth, Food, Sex & Death* (London: Pan Macmillan, 2020).

18 Robin Zheng, 'Why Yellow Fever Isn't Flattering: A Case Against Racial Fetishes', *Journal of the American Philosophical Association* 2, no. 3 (2016): 400–19; Sue K. Jewell, *From Mammy to Miss America and Beyond: Cultural Images and the Shaping of U.S. Social Policy* (New York: Routledge, 1993).

19 Ben Hunte, 'Grindr Fails to Remove Ethnicity Filter After Pledge to Do So', *BBC News* (26 June 2020).

20 Zheng, 'Why Yellow Fever Isn't Flattering'.

21 Ibid.

9 Never Love a Robot: Romantic Companions and the Principle of Transparency

1 Peter Berger and Hansfried Kellner, 'Marriage and the Construction of Reality: An Exercise in the Microsociology of Knowledge', *Diogenes* 12, no. 46 (1964): 1–24; Judith A. Hall and Shelley E. Taylor, 'When Love Is Blind: Maintaining Idealized Images of One's Spouse', *Human Relations* 29, no. 8 (1976): 751–61.

2 Ronald Noë and Peter Hammerstein, 'Biological Markets: Supply and Demand Determine the Effect of Partner Choice in Cooperation, Mutualism and Mating', *Behavioral Ecology and Sociobiology* 35, no. 1 (1994): 1–11; Hanna Kokko and Rufus A. Johnstone, 'Why Is Mutual Mate Choice Not the Norm? Operational Sex Ratios, Sex Roles and the Evolution of Sexually Dimorphic and Monomorphic Signalling', *Philosophical Transactions of the Royal Society of London, Series B: Biological Sciences* 357, no. 1419 (2002): 319–30.

3 Margaret Boden, Joanna J. Bryson, Darwin Caldwell, Kerstin Dautenhahn, Lilian Edwards, Sarah Kember, Paul Newman, Geoff Pegman, Tom Rodden, Tom Sorell, Mick Wallis, Blay Whitby, and Alan Winfield, 'Principles of Robotics' (2011). https://epsrc. ukri.org/research/ourportfolio/themes/engineering/activities/ principlesofrobotics/; see also by the same authors, 'Principles of Robotics: Regulating Robots in the Real World', *Connection Science* 29, no. 2 (2017): 124–9.

4 Joanna J. Bryson, 'The Meaning of the EPSRC Principles of Robotics', *Connection Science* 29, no. 2 (2017): 130–6.

5 Joan Roughgarden, Meeko Oishi, and Erol Akçay, 'Reproductive Social Behavior: Cooperative Games to Replace Sexual Selection', *Science* 311, no. 5763 (2006): 965–9.

6 Karen L. Kramer and Andrew F. Russell, 'Kin-Selected Cooperation without Lifetime Monogamy: Human Insights and Animal Implications', *Trends in Ecology & Evolution* 29, no. 11 (2014): 600–6.

10 Can Robots Be Moral Agents?

1 James H. Moor, 'The Nature, Importance, and Difficulty of Machine Ethics', *IEEE Intelligent Systems* 21 (2006): 18–21, https://doi.org/ 10.1109/MIS.2006.80.

2 Amanda Sharkey, 'Should We Welcome Robot Teachers?' *Ethics and Information Technology* 18 (2016): 283–29.

3 Amanda Sharkey and Noel E. Sharkey, 'Granny and the Robots: Ethical Issues in Robot Care for the Elderly', *Ethics and Information Technology* 14, no. 1 (2012): 27–40.

4 Noel E. Sharkey, 'The Evitability of Autonomous Robot Warfare', *International Review of the Red Cross* 94, no. 886 (2012): 787–99.

5 Ronald C. Arkin, *Governing Lethal Behavior in Autonomous Robots* (Boca Raton, FL: CRC Press, 2009).

6 Bryan Glick, 'Vilified then Vindicated: Victory for Sub-Postmasters in Post Office Trial Shows Risk of Tech Hubris', *Computer Weekly* (11 December 2019).

7 Noel E. Sharkey and Tom Ziemke, 'Mechanistic versus Phenomenal Embodiment: Can Robot Embodiment Lead to Strong AI?' *Cognitive Systems Research* 2, no. 4 (2001): 251–62.

8 Patricia S. Churchland, *Braintrust: What Neuroscience Tells Us about Morality* (Princeton, NJ: Princeton University Press, 2011).

9 David J. Gunkel, 'The Other Question: Can and Should Robots Have Rights?' *Ethics and Information Technology* 20 (2018): 87–99.

10 John Danaher, 'Welcoming Robots into the Moral Circle: A Defence of Ethical Behaviourism', *Science and Engineering Ethics* 26, no. 4 (2020): 2023–49.

11 Amanda Sharkey, 'Can We Program or Train Robots to Be Good?' *Ethics and Information Technology* 22 (2020): 283–95, https://doi.org/10.1007/s10676-017-9425-5.

12 Isaac Asimov, 'Runaround', *Astounding Science Fiction* (March 1942).

13 David Silver, Julian Schrittwieser, Karen Simonyan, Ioannis Antonoglou, Aja Huang, Arthur Guez, Thomas Hubert, Lucas Baker, Matthew Lai, Adrian Bolton et al., 'Mastering the Game of Go without Human Knowledge', *Nature* 550 (2017): 354–9, https://doi.org/10.1038/nature24270.

14 Sharkey, 'Can We Program or Train Robots to Be Good?', 283–95.

11 Words to Life: Creativity in Action

1 Pan Wang, Danling Peng, Ling Li, Liuqing Chen, Chao Wu, Xiaoyi Wang, Peter Childs, and Yike Guo, 'Human-in-the-Loop Design with Machine Learning', *Proceedings of the Design Society: International*

Conference on Engineering Design 1, no. 1 (2019): 2577–86, http://dx.doi.org/10.1017/dsi.2019.264. Reviewers' Choice Award, Design Society Distinguished Paper Award, www.cambridge.org/core/journals/proceedings-of-the-international-conference-on-engineering-design/article/humanintheloop-design-with-machine-learning/3A5B1A14E4F0701B3376F3C3AEB89D86.

12 An Anthropology and Philosophy of the Future

1 Aifric Campbell, *Scarlett and Gurl*, this volume.
2 Richard Watson, *Future Files: A Brief History of the Next 50 Years* (London: Scribe, 2007).
3 Alvin Toffler, *Future Shock* (London: Random House, 1970).
4 Richard Watson, *Digital vs Human: How We'll Live, Love, and Think in the Future* (London: Scribe, 2016).

Contributor Biographies

Ronny Bogani recently returned to academia following a twenty-year career practising trial law in the fields of product liability, consumer rights, criminal prosecution, immigration, and freedom of expression. He received his LLM in human rights law from the University of Edinburgh, an MRes in computer science from the University of Bath and is now pursuing his doctoral studies in artificial intelligence secure and explainable by construction (AISEC) at the University of Edinburgh. Ronny remains engaged in children's rights advocacy and policy development, and has participated in policy formation and development as a UN local rapporteur and workshop stakeholder in conjunction with the Vatican, Finland, and UNICEF, among others.

Joanna J. Bryson is a transdisciplinary researcher on the structure and dynamics of human- and animal-like intelligence. Her research ranges from systems engineering of artificial intelligence (AI), through autonomy, cognition, robot ethics, human cooperation, on to technology policy, and her research has appeared in Reddit and *Science*. She holds degrees in psychology from Chicago and Edinburgh, and AI from Edinburgh and MIT. She has additional professional research experience from Princeton, Oxford, Harvard, and LEGO, and technical experience in Chicago's financial industry, and international management consultancy. Bryson is the professor of ethics and technology at Hertie School of Governance in Berlin.

Julie Carpenter, PhD, is a research scientist based in San Francisco and a research fellow in California Polytechnic State University's Ethics + Emerging Sciences Group, a non-partisan organisation focused on the risk, ethical, and social impact of emerging sciences and technologies. Her research situates human–technology experiences within their larger cultural contexts and social systems to offer a framework for describing what phenomena are occurring and explain how people's cultural concepts and expectations, behaviours, and ideas adapt and change over time as they work and live with emerging technologies. In her work, Carpenter typically uses ethnographic research methods to explore sociocultural influences on human behaviours that inform topics such as trust, decision-making, and emotional attachment to artificial systems.

Stephen Cave is executive director of the Leverhulme Centre for the Future of Intelligence, where he leads a team of researchers across five programmes on the nature and impact of AI. He is also senior research associate in the Faculty of Philosophy and fellow of Hughes Hall (College) at the University of Cambridge. His own research is in the philosophy of technology; in particular, critical perspectives on AI, robotics, and life-extension technologies. Previously, Stephen earned a PhD in philosophy from Cambridge, and served as a British diplomat. He is the author of *Immortality* (Penguin Random House, 2012) and co-editor of *AI Narratives* (Oxford, 2020).

Anita Chandran is a writer and PhD student in ultrafast lasers and non-linear optics at Imperial College London. Her writing explores themes of science, sexuality, and race, and includes the award-winning short story '(Nothing But) Art', which depicts the romantic relationship between an AI and a visual artist. As a woman of colour in science, technology, engineering, and mathematics (STEM), she is interested in the importance of representation and role models for young people, specifically through the medium of fiction. She is the co-founder and fiction editor of *Tamarind*, a literary magazine focusing on the intersections of the arts and sciences.

Peter R. N. Childs is the professorial lead in engineering design at the Dyson School of Design Engineering, Imperial College London and chairperson at BladeBUG Ltd. His professional interests include creativity tools; design, flow, and heat transfer; sustainable energy; and robotics. Former roles include director of the Rolls-Royce University Technology Centre for Aero-Thermal Systems, director of InQbate, professor at Sussex University, founding head of the Dyson School of Design Engineering at Imperial, and chairperson at Q-Bot Ltd. He is professor at large for innovation design engineering at Imperial

College and the Royal College of Art, and professor of excellence at MD-H, Berlin. He is a fellow of the Royal Academy of Engineering.

Kate Devlin is a senior lecturer in the Department of Digital Humanities, King's College London. Her research in human–computer interaction and AI investigates how people interact with and react to technologies, both past and future. She is the author of *Turned On: Science, Sex and Robots* (Bloomsbury, 2018), which examines the ethical and social implications of technology and intimacy.

Kanta Dihal is a senior research fellow at the Leverhulme Centre for the Future of Intelligence, University of Cambridge. Her work intersects the fields of science communication, literature and science, and science fiction. She leads two research projects, Global AI Narratives and Decolonizing AI, in which she focuses on the portrayals and perceptions of AI across cultures. She is co-editor of *AI Narratives* (Oxford, 2020) and has advised the World Economic Forum, the UK House of Lords, and the United Nations. She obtained her DPhil on the communication of quantum physics at Oxford in 2018.

Mary Flanagan is a scholar of digital culture and an artist, with works exhibited at museums and galleries around the world such as the Whitney, the Guggenheim, Tate Britain, and other global institutions. In 2018, Dr Flanagan won the Award of Distinction at Prix Ars Electronica and is currently building an AI. She has held numerous fellowships, has published six books including *Critical Play* (MIT Press, 2013), and is distinguished professor and chair of film and media studies at Dartmouth College, USA.

Margaret Rhee is a poet, scholar, and new media artist. Her debut poetry collection, *Love, Robot*, was named a 2017 Best Book of Poetry by *Entropy Magazine* and awarded a 2018 Elgin Award by the Science Fiction Poetry Association and the 2019 Best Book Award in Poetry by the Asian American Studies Association. Her monograph, *How We Became Human: Race, Robots, and the Asian American Body*, is under review. She received her PhD from UC Berkeley in ethnic and new media studies and is assistant professor in the Department of Media Study at SUNY Buffalo.

Amanda Sharkey is a retired senior lecturer in computer science at the University of Sheffield, a member of Sheffield Robotics, and on the executive board of the Foundation for Responsible Robotics. Her background is interdisciplinary. After a first degree in psychology, she held research positions at the University of Exeter, MRC Cognitive Development Unit, Yale, and Stanford, USA. She completed a PhD in psycholinguistics at the University of Essex and conducted research in machine learning at the universities of Exeter and Sheffield. Her current research interests are in robot ethics, particularly the ethics of robot care for children and older people.

Roberto Trotta is a professor of astrostatistics at Imperial College London, currently on leave of absence to the International School of Advanced Studies in Trieste, Italy, where he is part of the senior leadership team establishing a new Data Science Institute. He is also a visiting professor of cosmology at Gresham College, London. His research focuses on cosmology, machine learning, and data science. An award-winning author and science communicator, he is the recipient of the Annie Maunder Medal 2020 of the Royal Astronomical Society for his public engagement work.

E. R. Truitt teaches in the Department of History and Sociology of Science at the University of Pennsylvania. Her first book, *Medieval Robots: Mechanism, Magic, Nature, and Art* (University of Pennsylvania Press, 2015), explored the rich history of artificial people and animals between 800 and 1450. She has published scholarly articles on the history of automata, clockwork, astral science, *materia medica*, courtly technology, and pre-modern concepts of AI, and has written for a broader readership in *Aeon*, *History Today*, and the *TLS*. She is currently finishing a book about the importance of experimental technology and historical periodisation in the history of science.

Richard Watson is futurist-in-residence at the Centre for Languages, Culture and Communication (CLCC) at Imperial College London, and has also worked with the Technology Foresight Practice at the

College. He has written five books, including *Digital vs Human* (Scribe, 2017), *Futurevision* (Scribe, 2010), and *Future Files* (Scribe, 2007), and contributed to many other works, including *Future Frontiers: Education for an AI World* (Melbourne University Press, 2017), *After Shock* (John August Media, 2020), and *A Street Through Time* (Dorling Kindersley, 2020).

Acknowledgements

Scarlett and Gurl started life as an imagining of the future of human love, but when it was finished I realised a novel would not be enough. I am so very grateful to the cross-disciplinary contributor group for their wonderful essays and their commitment to and support of *The Love Makers*' mission of public engagement in AI. Thanks to Holger Krapp and Peter Childs at Imperial College London for inspiration and detail on blowflies, aeronautics, engineering, and wearables. Holger's incredible blowflies are featured in David Attenborough's *Conquest of the Skies* (2015). Gratitude also to Alex Adamou at the London Mathematics Laboratory for a critical problem-solving lunch; Murray Shanahan at Imperial College London/Deep Mind for very useful conversation; Bas Heijne for sharing AI insights from his television series *The Perfect Human Being* (2015); Florence Noiville for citations from her interview with Don DeLillo; Megan Lalla-Hamblin for introducing me to Easter eggs and GitHub; Neil Sayers for his work consulting on programming and its representation; John Johnston at the London Mathematical Society; and Eileen McConnell for line-dance instruction. David Nutt's research on alcohol synthetics and Matthew Fisher's on salamanders was very helpful. For sound advice, huge thanks to Pete Ayrton who first published my writing; Sarah Kember and all at Goldsmiths Press for their enthusiasm; Lily McCraith and Ilyanna Kerr for cover design; Charlie Mounter for superb editing and advice; filmmaker Cal Murphy Barton for creative excellence; my colleagues in Room S310 – the banter hub of ideas; and the STEM students at Imperial College who pitch up in my classes to get black on white.

I am very grateful to the Museum of Fine Arts, Houston, and the Brown Foundation for my residency in 2018 at the Dora Maar House where this book finally found its way. Great adventures have their dark moments, so thanks to the supporters who kept the faith when I lost it – you know who you are!

Quotations

H. G. Wells for 'This black ignorance at our very feet' and 'All this world is heavy with the promise of greater things', from '*The Discovery of the Future:* A Discourse Delivered to the Royal Institution', 24 January 1902, www.telelib.com/authors/W/WellsHerbertGeorge/ prose/discoveryofthefuture/dicoveryofthefuture001.html.

The Forestry Commission, England, for 'A sharp axe and a cold heart', paraphrased from www.friendsofbournewoods.org.uk/wp-content/uploads/2017/04/Bourne-Wood-Booklet-Web.pdf.

Ayn Rand for 'To love a thing is to know and love its nature', 'joyous sense of confidence when looking at machines ... every part of the motors was an embodied answer to "Why?" and "What for?"' and 'All work is an act of philosophy', from *Atlas Shrugged* (1957).

Bertrand Russell for 'All the conditions of happiness are realised in the life of the man of science', paraphrased from *The Conquest of Happiness* (1930).

Sigmund Freud for 'work and love, love and work', from '*Lieben und arbeiten*', which is one of the many unsourced quotes attributed to Freud.

Virginia Woolf for 'Every woman should have a room of her own and five hundred pounds', paraphrased from *A Room of One's Own* (1929), where the room and the money are mentioned in different parts.

Karel Čapek for 'You know where the word robot comes from, Gurl? *Robota*, it's Czech.' The word was first used by Karel Čapek in his play *R.U.R. (Rossum's Universal Robots)*, which premiered 25 January 1921 at the National Theatre in Prague, http://preprints.readingroo. ms/RUR/rur.pdf, translated by Paul Selver and Nigel Playfair.

Ralph Waldo Emerson for 'The only way to have a friend is to be one', from *Essays, First Series* (1841).

Finally, thanks to Ray Bradbury for good company on my journey: 'Duplicate self or friends; new humanoid plastic 1990 models, guaranteed against all physical wear ... From $7,600 to our $15,000 deluxe model ... Marionettes, Inc., is two years old and has a fine record of satisfied customers behind it. Our motto is "No Strings Attached"' (Ray Bradbury, 'Marionettes Inc.' *The Illustrated Man*, 1951).